GENE CONSERVATION AND EXPLOITATION

STADLER GENETICS SYMPOSIA SERIES

CHROMOSOME STRUCTURE AND FUNCTION
Impact of New Concepts
Edited by J. Perry Gustafson and R. Appels

GENE CONSERVATION AND EXPLOITATION
Edited by J. Perry Gustafson, R. Appels, and P. Raven

GENE MANIPULATION IN PLANT IMPROVEMENT
Edited by J. Perry Gustafson

GENE MANIPULATION IN PLANT IMPROVEMENT II
Edited by J. Perry Gustafson

GENETICS, DEVELOPMENT, AND EVOLUTION
Edited by J. Perry Gustafson, G. Ledyard Stebbins, and Francisco J. Ayala

GENE CONSERVATION AND EXPLOITATION

20th Stadler Genetics Symposium

Edited by

J. Perry Gustafson

USDA-ARS
University of Missouri
Columbia, Missouri

R. Appels

CSIRO
Canberra, ACT, Australia

and

P. Raven

Missouri Botanical Gardens
St. Louis, Missouri

PLENUM PRESS • NEW YORK AND LONDON

Library of Congress Cataloging in Publication Data

Stadler Genetics Symposium (20th: 1991: University of Missouri–Columbia)
Gene conservation and exploitation: 20th Stadler Genetics Symposium / edited by J. Perry Gustafson, R. Appels, and P. Raven.
p. cm.—(Stadler genetics symposia series)
"Proceedings of the Twentieth Stadler Genetics Symposium . . . held September 30–October 1, 1991, at the University of Missouri, Columbia, Missouri"—T.p. verso.
Includes bibliographical references and index.
ISBN 0-306-44533-6
1. Germplasm resources, Plant—Congresses. 2. Germplasm resources, Plant—Utilization—Congresses. I. Gustafson, J. P. II. Appels, R. III. Raven, P. IV. Title. V. Series.
SB123.3.S83 1991 93-23116
631.5′23—dc20 CIP

Proceedings of the Twentieth Stadler Genetics Symposium, Gene Conservation and Exploitation, held September 30–October 1, 1991, at the University of Missouri, Columbia, Missouri

ISBN 0-306-44533-6

A Division of Plenum Publishing Corporation
233 Spring Street, New York, N.Y. 10013

Printed in the United States of America

The editors would like to dedicate the 20th Stadler Genetics Symposium to **Dr. Gyorgi Rédei** for his tireless effort in organizing the first thirteen Stadler Symposia.

ACKNOWLEDGMENT

The editors gratefully acknowledge the generous support of the following contributors: College of Agriculture, Department of Agronomy, Plant Science Unit, Division of Biological Sciences, Graduate School and School of Medicine, University of Missouri-Columbia; Agracetus, ICI Seeds, Monsanto Company; Northrup King Company; and Pioneer Hi-Bred International, Inc., who helped make the 20th Stadler Genetics Symposium a success.

The speakers, who spent a tremendous amount of time preparing their manuscripts and lectures, are gratefully acknowledged. Without their expertise and dedication the Symposium could not have taken place.

We wish to thank the local chairpersons for their effort to see that everyone in the respective sessions were well taken care of during the Symposium.

The behind-the-scene and on-site preparation was excellently handled by Joy Williams from Conferences and Specialized Courses, University of Missouri, who tirelessly handled all of our peculiar requirements and made sure everything was extremely well organized.

Many thanks are due to Joyce Reinbott, University of Missouri, for her excellent secretarial help in handling all the correspondence and typing. A special thanks from JPG goes to Kate Houchins and Kathleen Ross for keeping the lab running.

J.P. Gustafson, R. Appels, and P. Raven

June 03, 1993
Columbia, Missouri
Canberra, ACT, Australia
St. Louis, Missouri

CONTENTS

BROADENING THE BASIS OF PLANT RESOURCE CONSERVATION

Vernon H. Heywood

Chief Scientist, Plant Conservation
IUCN - The WorldConservation Union
Descanso House
199 Kew Road
Richmond
Surrey, TW9 3BW
U.K.

INTRODUCTION

Worldwide there are many thousands of species of higher plants which are used by man and can therefore be considered resources (Heywood 1991). About 5,000 species have been cultivated at one time or another and there are many more which are harvested or used directly from the wild. As an illustration of this, the Plant Resources Project of South-East Asia records over 8,000 species in its basic list of species used by mankind in that area (Lemmens et al., 1989) and assuming similar percentages of the floras of other tropical regions are similarly employed, then we can extrapolate to a figure of 25-30,000 species for the tropics as a whole. In addition several thousand species are used by Man in the temperate regions of the world. Up to 25,000 species have been used in herbal medicines and many thousand species of ornamentals are grown in parks and in public and private gardens and as street trees.

Very few of these species enter into world trade and only about 20-30 of them are considered staple crops that supply most of human nutrition needs although a recent study by Prescott-Allen and Prescott-Allen (1990) based on supply data indicates that a total of 103 species contribute 90% of national per capita supplies of food plants. Not surprisingly, in the conservation of plant genetic resources most attention has been paid to such food crops and to industrial crops such as cacao, coffee, rubber, oil palm, jute, cotton, tobacco, sugar cane, as well as forages, green manures and timber trees.

Despite various calls in recent years for a broadening of the basis of

Gene Conservation and Exploitation
Edited by J.P. Gustafson *et al.*, Plenum Press, New York, 1993

plant genetic resource conservation, there are few signs of a coordinated approach to the conservation of plant resources, either at national or at international level, that includes the broad range of known resources such as medicinal species, indigenous crops, ornamentals and agroforestry species, as well as the major crop, forage and timber species, and which involves all the available institutional resources such as arboreta and botanic gardens as well as crop germplasm centres. National germplasm systems are usually restricted to crop and forestry species and the recently published report on the US National Germplasm System (NRC 1991), for example, devotes little attention to the needs of wild species (apart from some crop relatives) or to institutional resources such as botanic gardens which deal with such plants. On the other hand, the mandate of the Botanic Gardens Conservation Secretariat, which comprises an international network of botanic gardens concerned with conservation, includes coordinating the efforts of the member gardens to collect and conserve wild germplasm of wild plants, especially those that are rare or endangered, and with special emphasis on species of economic value or potential, including medicinal plants (WWF/IUCN/BGCS 1989). The role of botanic gardens in germplasm conservation is discussed in detail below. Another initiative is Project Noah launched by the US Agency for International Development (USAID) as part of its international biodiversity activities. Its aim is to act as an international rescue mission for the thousands of animals and plant species faced with the prospect of imminent extinction and its mandate includes establishing an AID policy and a strategic plan for the *ex situ* conservation of crop plants, threatened wild plants, domesticated and wild animals and useful, but unexploited wild plant species and forest species. Just how effective Project Noah will be must depend largely on the level of funding that is allocated.

PLANT RESOURCES VS GENETIC RESOURCES

The term plant genetic resources is often restricted by plant breeders and those working in the genetic resources sector to the few hundred crop, forage and forestry species that they work with and whose genetic diversity can be used for plant domestication and improvement. It may thus be defined as "the economic, scientific or societal value of the heritable materials contained within and among species" (FAO 1989). Since in the vast majority of cases, little, if anything is known of the heritable features of the majority of the tens of thousands of species used by Man, they cannot in the strict sense be regarded as plant genetic resources but they certainly must be considered resources and to ignore them in our considerations of germplasm conservation is surely not acceptable. A broader approach is taken by Oldfield

(1989) in her valuable survey "The value of Conserving Genetic Resources" which covers both wild and domesticated species.

In view of the very large numbers of species that must be considered as plant resources, not to mention an incalculable number of species that are of scientific or ecological importance, it is quite evident that we need to examine closely what are the possibilities of taking effective action to conserve samples of their germplasm, either *in situ* or *ex situ* or any combination of these, while they are still available, especially as many of them are at risk of extinction. Clearly a system of priorities has to be devised and it is surprising that so little attention has been paid to this issue. There seems to be an assumption in most writing on biodiversity conservation that the conservation of habitats will ensure the survival of the tens of thousands of plant species known to be at risk and the tens of thousands more that will become threatened as their habitats decline or are eliminated. Indeed much play has been made recently of the need to give priority in the conservation of plant (and animal) biodiversity and resources to *in situ* approaches. Others, however, draw attention to the erroneous assumption that *ex situ* conservation activities are already being fully attended through bodies such as the commodity-based International Agricultural Research Centres (IARCs) of the CGIAR, the International Board for Plant Genetic Resources (IBPGR) and the Food and Agricultural Organization (FAO) of the United Nations (cf. Cohen et al 1991) and call for priority to be given to broadening *ex situ* conservation action.

This paper will explore just how feasible it is in practical terms to conserve large numbers of targeted plant species and proposals will be made for both national and international strategies to address these issues.

THE BIODIVERSITY CONTEXT

The conservation of plant genetic resources has to be a viewed within the context of what is happening to biodiversity as a whole. Much has been written in the past few years about the magnitude of the crisis facing global biodiversity (e.g. Wilson and Peter 1988). Few today would dispute the prediction that, despite many calls for their conservation, a large proportion of the world's remaining forests and other ecosystems, especially in the tropics, are destined to disappear over the coming decades through conversion for other uses, largely in response to the needs of increasing human populations.

A major consequence of this loss of habitat that we are witnessing in most of the world's terrestrial ecosystems is that tens of thousands of species are suffering contractions of their range and consequent loss of genetic variation which in some cases may lead eventually to their local or even total extinction sometime over the coming decades. Assessments differ as to the numbers of species that will become totally extinct but what is certain is that tens of thousands will be reduced to small populations in fragmented habitats before they finally become extinct and many of them will hang on in such vegetation fragments for decades, if not longer, so that there will be time to undertake conservation action if desired (see papers in Saunders et al 1987). This presents a quite different scenario from the one that is normally postulated of massive extinctions within the coming 20-40 years, and therefore poses a challenge of awesome proportions for conservation.

The total extinction scenario, is at the same time more tragic and also neater in that once extinct, species cannot be reestablished except through reintroductions in a limited number of cases. What is surprising is that in neither case has the conservation community considered the consequences seriously: faced with the likely partial or total loss of tens of thousands of plant species over the next century, no global rescue mission has been attempted or even envisaged, with the exception of the initiatives of BGCS and Operation Noah mentioned above. It would seem as though the enormity of the tragedy confronting us has either not been appreciated or has been set aside because it is too large to comprehend. No-one has suggested that a loss of plant diversity on this scale does not matter: indeed much of what has been written about the biodiversity crisis has drawn attention to the probable losses of plants of potential economic importance and the sources of new drugs, crops etc. Yet with the exception of the valiant efforts of institutions such as Missouri Botanical Garden, New York and Kew and many institutions in the tropics, to attempt to complete the inventory of plant diversity before the forests and other habitats are cleared, the normal reaction has been to call for an increase in the world's protected area network. However, such an approach, which I would strongly support, is only one part of the solution and does nothing for those species whose loss is imminent which do not occur in the areas suitable for protection. Even for those species that do occur in actual or yet-to-be-declared protected areas, their survival is not necessarily guaranteed by such ecosystem conservation as is discussed below.

CONSERVATION STRATEGIES

It is generally accepted that, others things being equal, conservation of species and populations in their natural habitats (*in situ*) so that their populations can continue to evolve under natural conditions is to be preferred over any other approach. Today the conservation of biodiversity is, as Soule (1991) comments, virtually equivalent to the protection of bounded wildlands with relatively little human disturbance.

The advantages of such *in situ* conservation are well rehearsed and will not be repeated here. Such an approach has, however, severe limitations and the future of many such *in situ* reserves will become less secure in the face of biogeographic, social and political factors and increasingly less attractive forms of conservation will come to play a dominant role.

As we have noted above *ex situ* conservation has been used as a major strategy to safeguard genetic resources of species of known present-day value (FAO 1991), i.e. crop plants, forest trees and forage species, and we will have to consider how far such an approach can be employed for a wide range of species, whether they are of economic interest or not, by botanic gardens and other institutions.

What we advocate in the *Botanic Gardens Conservation Strategy* is the adoption of an integrated approach to conservation in the sense that the techniques to be applied should be the most appropriate to the particular circumstances (WWF/IUCN/BGCS 1990). Soule (1991) in fact describes eight paths to biotic survival: *in situ*, *inter situ*, extractive reserves, ecological restoration projects, zooparks, agroecosystems and agroforestry projects, living *ex situ* programmes and suspended *ex situ* programmes. Although there are many other schemes outlining conservation approaches (cf. McNeely et al., 1990), Soule's list gives an idea of the diversity of techniques open to us. As Falk (1990) notes, the use of a wide range of conservation methods is compelled by the diversity in objects of conservation and threats that they encounter. We have moved on from the earlier *laissez faire* ethic that dominated conservation philosophy in earlier decades towards acceptance of a much more interventionist approach on the one hand and the acceptance of a broader range of options and techniques for conservation that are now available to us on the other hand.

In situ conservation

We do not live in an ideal world and we have to face the fact that few large areas of intact vegetation are likely to remain in the future, except in a few favoured localities, apart from those that we are able to maintain and manage effectively. It is important to distinguish, therefore between the following types of *in situ* situation (1) the conservation of biodiversity in protected areas subjected to various degrees of human intervention, (2) the conservation of targeted species in protected areas and (3) conservation of targeted species in smaller, often fragmented pieces of vegetation or in secondary communities.

(1) Conservation of biodiversity in protected areas. In the conservation of biodiversity through ecosystem or landscape conservation (as in national

parks, biosphere reserves and other kinds of protected areas), the aim of management is to preserve their integrity as samples of functioning ecosystems or landscapes and not to conserve any particular species, populations or genes they contain. Indeed the areas are often not strategically sited on the basis of genetic or taxonomic criteria (FAO 1991). It is obvious that many tens of thousands of species will be conserved to some degree through the conservation of the habitats in which they occur but this unfocussed approach provides no guarantee that any particular species will be adequately conserved. It is possible to conserve an ecosystem and still lose individual species or to save the species and lose genetically distinct populations (Wilcox 1982).

(2) Conservation of targeted species in protected areas. A distinction has to be made between the conservation of biological diversity through ecosystem or landscape conservation (as in national parks, biosphere reserves and other kinds of protected areas) and the conservation of clearly targeted plant or animal species in networks of special conservation areas such as genetic reserves, gene sanctuaries, wildlife reserves and the like. In the former case, as we have seen, the areas are managed so as to maintain their integrity as samples of functioning ecosystems or landscapes and not to preserve any particular species, populations or genes which they contain but rather in a general way to avoid the loss of species within them. In the latter case the reserve areas are dynamic units of conservation for the range of genetic variability in the populations of particular species they house that are of actual or potential economic importance or of scientific or other special value, and managed for this purpose.

The role and effectiveness of *in situ* conservation for the conservation of species and populations will depend on the degree of management applied (Lleras 1991). The maintenance of a site in its natural state will not necessarily conserve genetic variability in any particular target species and its populations. *In situ* conservation for this purpose requires carefully planned surveying and systematic management so that the challenge is not so much to set aside and guard protected areas for *in situ* conservation of species as such as to maintain genetic variability in the target species within a mosaic of economically and socially acceptable land use options (FAO 1991).

A high degree of intervention in an area may be compatible with the conservation of a target species within it. In the much cited Sierra de Manantlán Biosphere reserve in Mexico, its flagship species *Zea diploperennis* (locally known as *milpilla* an important relative of maize, containing valuable germplasm for disease resistance, occurs only in or near human agricultural habitations and may actually grow within the actual maize field itself in local *coamil* slash and burn agricultural systems. Indeed it is planned to continue the *coamil* system in one or two highly restricted areas within the Biosphere reserve so that the *milpilla* populations will remain healthy and safe from extinction (Guzman and Iltis 1991). As Guzman and Iltis comment, it is paradoxical that the flagship species, *milpilla*, to which the enormous 140,000 ha Sierra de Manantlán Biosphere reserve owes its initiation has turned out to be favoured by human disturbance.

Various levels of management may be applied to *in situ* conservation systems varying from no specific management but only monitoring of the populations of the species of conservation concern within the area under question to ensure that they are being maintained, all the way through to intensive management (cf Lleras 1991). Likewise the protected area concerned may be subjected to a various levels of human intervention and use.

Although the principles of *in situ* conservation of genetic resources have been well established (cf. Wilcox 1984; Ingram 1984; FAO 1991), few such areas have been established so far and consequently little practical experience of maintaining them has been acquired so far. Examples of *in situ* reserves are the gene sanctuaries in the Garo hills of India for wild relatives of *Citrus* and similar reserves for fruit trees in other parts of India, in China and the USSR. There are several orchid sanctuaries in India, conservation areas for wild coffee in Ethiopia and a reserve for wild chilies (*Capsicum annuum* var *aviculare*) in the Coronado National Forest in Arizona, USA. The Brazilian National Centre for Research in Genetic Resources (CENARGEN) has established seven genetic reserves for a variety of economically important species and is planning a further five (Lleras 1991; L.Coradin personal communication). In addition CENARGEN is planning to establish seedling banks of tropical perennials *in situ.*

We do not have more than an approximate idea of the number of species that will require some form of managed *in situ* conservation. A useful review of the *in situ* conservation needs for the wild relatives of crop species was prepared by Prescott-Allen for the FAO Commission on Plant Genetic Resources (FAO 1985) but no attempt has been made to make a similar survey of the very large number of other species of economic or scientific interest whose *in situ* conservation is becoming necessary.

A wide array of actions is needed to carry out a programme of *in situ* conservation including survey and inventory of the areas, documentation and monitoring, identification of target species, setting priorities both in terms of species and areas, ecological and ethnobiological studies and establishing training needs. It has to be recognized that no serious attempt has been made to treat *in situ* conservation of targeted species as a major item on the conservation and development agenda, despite frequent attempts to highlight its importance. If we are to conserve a much broader array of plant genetic resources than at present, then a concerted effort needs to be planned and implemented in this area.

(3) Conservation of targeted species in vegetation fragments. A very large number of species are now reduced to small populations whose survival in the medium to long term cannot be assured without habitat farming or some other form of intervention. Such species often occur in fragments of ecosystems which have otherwise largely disappeared.

This kind of situation is becoming increasingly common in most parts of the world. In Europe, for example, very little primary native vegetation exists after thousands of years of deforestation, pasturage, agriculture, industrial and commercial development: much of the characteristic vegetation of the Mediterranean basin is of secondary subseral communities caused or maintained by human intervention. Yet most of Europe's flora in terms of species continues to exist, with very few recorded extinctions, although many species have suffered severe contractions in range or loss of local races or ecotypes. Some of them are reduced to very small populations and their long term conservation requires the setting up of a series of mini-reserves, supported by *ex situ* conservation of seed or other material.

In Australia, although subjected to intensive development for only 200 years, 70% of the plant communities which existed in 1788 have been modified in some way and over 65% of the original tree cover has been removed, with up to 75% of the rainforest cleared for grazing and agriculture. Again despite this massive intervention by man, only 100 out of an estimated 25,000 plant species have been recorded as extinct in the past 200 years (Saunders et al 1990). In these fragments of vegetation the populations of plant species represented will inevitably be reduced in size, as we have seen, often to critical levels or facing imminent extinction although there is evidence that some species have a remarkable ability to survive in such small populations and maintain a considerable amount of genetic variability (cf Moran and Hopper 1987).

Island floras are also particularly vulnerable to habitat loss and to fragmentation of habitats and species populations. An example is the threatened flora of the island of Rodrigues where nine of the species are reduced to ten individuals in the wild, three to single individuals (Strahm 1989). In such circumstances there is clearly an appreciable risk that the remaining populations will not survive in the wild so that it is essential that both *in situ* and *ex situ* work be carried out in tandem. Any one approach on its own carries a serious risk of failure. Many similar situations occur on other islands both in the Indian Ocean and in other parts of the world.

This widespread fragmentation of habitats and species populations, which is an inevitable consequence of the massive contraction of ecosystems that we are witnessing around the world, poses an acutely difficult set of problems for conservation. The establishment and management of mini-reserves and supporting programmes of *ex situ* conservation for thousands, if not tens of thousands of species, has not hitherto been seriously considered although small individual conservation programmes for a few of them have been put in hand, as in the case of the Rodrigues endangered flora. To ignore this problem is to hasten the ultimate extinction of a substantial percentage of the world's flora.

Ex situ conservation

For the conservation of the genetic resources of the small number of plant species that supply most of the world's food needs, *ex situ* conservation has been the major technique employed. The crop genetic resources sector in particular depends heavily on the important networks of genebanks that house the germplasm needed for breeding programmes. The number of species that have been intensively and extensively sampled and stored in genebanks is about 500 (Hawkes 1987). These are held in the genebanks maintained by the CGIAR crop centres and by national genebanks such as the US National Germplasm System, the Nordic Gene Bank and the Vavilov Institute, USSR. The world's agricultural genebanks contain some 2.5 million accessions, most of which are crop cultivars, with only a small percentage referring to wild species. The US National Germplasm System is the world's largest distributor of germplasm, supplying more than 130,000 samples from its collections each year to more than 100 nations (NRC 1991) and currently contains more than 380,000 different accessions of some 8,700 species.

Although many agricultural genebanks do contain accessions of wild species, the numbers and degree of sampling is often inadequate. Moreover, unless there is some special interest in these wild species, as in the case of crop relatives, there is usually no programme of activity carried out on them. There is no overall strategy for the conservation of germplasm of wild species, either *in situ* or *ex situ*, nor is there an organization comparable to IBPGR with a mandate to cover this area of genetic resource conservation. The FAO Panel of Experts on Forest Gene Resource established in 1968 takes stock of the situation on forest genetic resources and advises FAO on priorities for action, by region and by species but the list of priority species, drawn up by the Panel, while extremely useful is incomplete. Few National Germplasm

Strategies take into account the needs for wild species conservation. A major, if largely unacknowledged, role is played by the world's botanic gardens and arboreta which house samples, albeit very small, of very large numbers of wild plant species. Indeed one of the major goals of Botanic Gardens Conservation International, first established by IUCN - The World Conservation Union in 1987 but now an independent although associated organization, is the mobilization of the efforts of botanic gardens and arboreta worldwide to conserve genetic stocks of wild species. Priority is given to species of economic importance and especially those that are rare or endangered.

THE BOTANIC GARDEN SYSTEM

There are currently over 1,500 botanic gardens and arboreta in the world (Heywood and Heywood 1991). The range of material that they contain is quite remarkable and represents the largest assemblage of plant diversity outside nature. Tens of thousands of species are represented in their collections although often by only one or a few individuals. Most of this material occurs as "living Collections" (as opposed to seed) either in the open or in greenhouses (Heywood 1990). Until recently most botanic collections were assembled for reasons other than conservation but in the last ten years or so many gardens have devoted efforts to building up conservation collections and today over 500 of them possess such collections.

An increasing number of gardens is developing national programmes for the conservation of locally or nationally endangered species with a view to their long term conservation, and, in some cases, reintroduction into the wild. This involves research into methods of cultivation, propagation, seed storage and germination, and the need for close cooperation with managers of protected areas and other conservation bodies. Examples of such programmes may be found in participating gardens of the Center for Plant Conservation, St Louis, which aims to ensure the conservation of all the endangered species of the flora of North America. Other examples include the Jardín Botánico Viera y Clavijo, Gran Canaria which holds important conservation collections for the Macaronesian region, several Australian botanic gardens such as Adelaide, Perth and Canberra which maintain collections of endangered Australian species, and the Jardim Botânico do Rio de Janeiro which is working on the endangered flora of the Mata Atlantica. The French Botanical Conservatoires at Brest, Porquerolles and Nancy are dedicated to the conservation of endangered French species, including those of the overseas Departments such as Réunion and Guadeloupe. Several Chinese botanic gardens are engaged in the conservation of rare and endangered species, including medicinal plants which are of major concern and importance in China (He, Yuan, Qin and

Dudley 1990; He and Chen 1991). Indeed many botanic gardens in different parts of the world are engaged in programmes of medicinal plant conservation (Heywood 1991). Extensive *ex situ* collections of forest species are grown in special areas by several Chinese arboreta (He, Yuan, Qin and Dudley 1990).

PLAN OF ACTION

As noted above, we need to mount a global effort to conserve populations of a substantial number of the species whose survival is threatened by destruction or loss of habitat. This will involve the following elements:

- Σ Inventory and listing of candidate species
- Σ Assessing priority species and areas
- Σ Devising appropriate sampling techniques
- Σ Arranging for samples of seed or other materials to be collected for *ex situ* conservation
- Σ Developing links between protected areas and botanic gardens and establishing minimum facility gardens or nurseries in association with protected areas so that integrated conservation policies can be adopted
- Σ Developing the infrastructure and technical facilities of botanic gardens as the main *ex situ* conservation facilities for wild species, in close association with the agricultural and forestry genebanks, IBPGR and FAO.
- Σ Establishing an information system for the conservation of germplasm of wild species, in association with Botanic Gardens Conservation International, the World Conservation Monitoring Centre, IUCN Species Survival Commission, FAO and IBPGR.
- Σ Creation of integrated conservation strategies for species populations restricted to vegetation fragments.
- Σ Assessment of costs and benefits.

CONCLUSIONS

The number of plant species that are in need of conservation action now and in the coming decades is several orders of magnitude greater than the number for which recovery plans are currently available or even envisaged. This will require a major rethinking of our conservation priorities. While many of these species will survive as the result of the conservation of the ecosystem or habitat in which they occur, for many others a more targeted and specifically managed programme of action will be needed. The large number of depleted species populations restricted to vegetation fragments pose special problems and will in most cases need integrated conservation action involving both *in situ* conservation and *ex situ* backup. Botanic gardens will play a key role in these actions and will become the leading resource centres for conservation of wild species and depositories for *ex situ* collections, both short term and long term.

REFERENCES

Cohen, J.L.,Williams, J.T.,Plucknett, D.L.and Shands, H.,1991,*Ex situ* conservation of plant genetic resources: global development and environmental concerns,*Science*,263:866-872.

Falk, D.A.,1990,Integrated strategies for conserving plant diversity,*Ann.Missouri Bot.Garden*,77:38-47.

FAO,1989,"Plant Genetic Resources.Their Conservation *In Situ* for Human Use,"FAO,Rome.

FAO,1991,"Strategies for the Establishment of a Network of In Situ Conservation Areas,"FAO,Rome.

Guzman, M.R.and Iltis, H.H.,1991,Biosphere reserve established in Mexico to protect rare maize relative, *Diversity*,7:82-84.

He, Shan-an and Cheng, Zhong-ming,1991,The role of Chinese botanical gardens in conservation of medicinal plants,*in:*"Conservation of Medicinal Plants",Akerele,O.,Heywood,V.and Synge,H.,eds., Cambridge University Press,Cambridge.

He, Shan-an, Yuan, Yi-Wei, Qin, Hui Zhen and Dudley, T.R.,eds.,1990,"Present Conservation Status of Rare and Endangered Species in Chinese Botanical Gardens,"Jiangsu Science and Technology Publishing House, Nanjing.

Heywood, V.H.,1990,Botanic gardens and the conservation of plant resources,*Impact of Science on Society*,158:121-132.

Heywood, V.H.,1991,Botanic gardens and the conservation of medicinal plants,in:"Conservation of Medicinal Plants",Akerele,O.,Heywood,V.and Synge,H.,eds.,Cambridge University Press, Cambridge.

Heywood, V.H.,1991,Conservation of germplasm of wild plant species,*in:*"Genetic Conservation and Sustainable Development,"Hindar,K.,Sandlund,O.T.and Brown,A.H.D.,eds., Universitetsforlaget,Oslo.

Heywood, C.and Heywood, V.H.,1991,"International Directory of Botanical Gardens V,"Koeltz Scientific Books, Koenigstein.

Lemmens, R.H.M.J., Jansen, P.C.M, Siemomsma, J.S.and Stavast, F.M.,1989,"Plant Resources of South-East Asia Basic List of Species and Commodity Grouping,Version 1,"PROSEA Project,Wageningen.

Lleras, E.,1991,Conservation of genetic resources *in situ,Diversity*,7:72-74.

McNeely, J.A.,Miller, K.R.,Reid, W.V.,Mittermeier, R.A.and Werner, T.B.,1990, "Conserving the World's Biological Diversity,"International Union for Conservation of Nature and Natural Resources, World Resources Institute, Conservation International, World Wildlife Fund US, and the World Bank,Gland, Baltimore and Philadelphia.

Moran, G.F.and Hopper, S.D.,1987,Geographic population structure of eucalypts and the conservation of their genetic resources,*in:*"Nature Conservation:the Role of Remnants,"Saunders, D.A., Arnold, G.W., Burbidge, A.A., and Hopkins, A.J.M., eds.,Surrey Beatty Pty Limited, Chipping Norton.

Oldfield, M.L.,1989,"The Value of Conserving Genetic Resources,"Sinauer Associates Inc.,Sunderland.

Prescott-Allen, R.and Prescott-Allen, C.,1990, How many plants feed the world?,*Conservation Biology* 4:365-374.

Saunders, D.A., Arnold, G.W., Burbidge, A.A. and Hopkins, A.J.M., eds., 1987, "Nature Conservation: the Role of Remnants of Native Vegetation,"Surrey Beatty & Sons, Chipping Norton.

Saunders, D.A. , Hopkins, A.J.M. and How,R.A.,eds.,1990,Australian Ecosystems:200 Years of Utilization, Degradation and Reconstruction,*Proc. Ecological Society Australia*,16.

Soule, M.E., 1991,Conservation:tactics for a constant crisis,*Science*,253:744-750.

Strahm, W., 1989,"Plant Red Data Book for Rodrigues,"Koeltz Scientific Books,Konigstein.

Wilcox, B.A., 1984, *In situ* conservation of genetic resources:determinants of minimum area requirements, *in:*"National Parks Conservation and Development. The Role of Protected Areas in Ssutaining Society,"McNeely, J.A. and Miller,K.R.,eds.,Smithsonian Institution Press, Washington DC.

Wilson, E.O. and Peter, F.M., eds., 1988, "Biodiversity, "National Academy Press, Washington, DC.

WWF/IUCN/BGCS, 1989, "The Botanic Gardens Conservation Strategy," WWF, IUCN, Gland and Richmond.

SCIENTIFIC ISSUES AFFECTING GENE CONSERVATION AND EXPLOITATION OF SOME TROPICAL PERENNIALS

John Trevor Williams

International Program for Tropical Tree Crops
Conservation and Development
International Fund for Agricultural Research
1611 N. Kent Street, Suite 600
Arlington, VA 22209

INTRODUCTION

Since environmental quality and population growth are moving in opposite directions, there is an urgent need to manage resources and to produce crops in a sustainable way. Resources to be managed include genetic resources of a wide array of plants which will be used to intensity production of food crops on existing agricultural lands, to rehabilitate degraded marginal lands, and to slow the destructive use of poor lands (Wilken, 1991). There is a resurgence of interest in the use of tropical tree crops for a variety of purposes, ranging from crop diversification to agroforestry, in the context of sustainable development. Unfortunately, there are a number of important scientific issues which are not being addressed adequately and one which more concentrated and coordinated research is necessary, especially in relation to gene conservation and exploitation.

The great strides made on collection, conservation and documentation of plant genetic resources over the past 25 years grew out of the results of the plant breeding successes in the 1960s which greatly enhanced production of food crops around the world. Responsibility was taken by the Consultative Group on International Agricultural Research to mobilize funding, and progressively genetic conservation collections came into being for a wide range of plant germplasm where the major methods of preservation were those of *ex situ* storage. In parallel, work on forest trees progressed, albeit at a slower rate, under the guidance of the Food and Agriculture Organization. Apart from a limited number of tree crops of the tropics -- cocoa, rubber and banana, for example -- where there was interest in uncoordinated national activities on a limited number of species (Smith *et al.*, in press).

The neglect stems from a feeling of little sense of urgency due to a widespread misunderstanding that the tropical tree crops are mainly plantation species grown by large estates which are a relic of past colonial eras, when the bulk are actually grown by small farmers. This is coupled with the mistaken belief that the available germplasm, though of limited variability, has been adequate for past needs and improved agronomic practices have generally been sufficient to sustain increased yields. Further, there is little public sector funding for genetic conservation work on woody crops and few advocacy groups in the public sector.

Interest has, however, increased due to pressures from the environmental conservation community and increasing recognition of the economic and social realities of development in tropical countries. Although it is now becoming much more widely appreciated that tropical tree species are capable of producing a greater range of products and in far more abundance than is currently the case, few have been fully domesticated and subjected to the selection processes and breeding improvements that have transformed the staple food crops. Many are genetically virtual wild species. This paper attempts to outline the research that needs to be done before the diversity of tropical tree crops can be more fully understood and the genetic diversity assessed and explored. An in-depth analysis of current knowledge must first be done before addressing the questions: what to sample; how to sample; how to conserve; and how to streamline utilization.

BACKGROUND

There are few relevant scientific studies on the population structure of specific tropical woody species. Whereas the variation patterns of temperate trees are fairly well understood and their reproductive systems are relatively easy to manipulate, this is not true for tropical species. Less than 20 years ago this was the case even for major timber species such as teak and *Eucalyptus*; most breeding programs for tropical trees relied on simple tree selection and did not progress much further (Bawa, 1976). For some timber species large areas of plantations were established, often with seed of doubtful origin and generally with little information on the genetic structure of the parent population (Wood, 1976). A number of tropical tree crop plants had also originated in the same way. Nonetheless, knowledge of population structures would have enabled improvement beyond selection and provenance trials which have remained the mainstay of tropical forest tree improvement.

Tree crops contrast with the timber trees in that, for those that have received breeding attention, past improvement efforts often relied on small plant introduction collections of a limited number of samples, and often single plants. These rarely represented populations nor had they been designed to encompass the spectrum of variation of the genepool.

To answer the questions related to genetic conservation, it is necessary to understand spatial and temporal aspects of the populations. Our current knowledge is outlined below, and thereafter scientific constraints will be discussed.

Populations in Space

The genetic resources of tropical tree crops are divided among wild populations, usually in the forests, sometimes specific populations adapted to major ecozones and primitively cultivated materials associated with habitation. Genetically such populations represent the landraces of arable crops and the pool of variation which provides sources of diversity for adaptation to new environments and potential materials for breeding. The primitively cultivated materials may be of no more special interest than the wild populations, especially in cases where over time semi-domesticates have periodically reverted to the wild.

In space, populations of a species may occupy vast continuous geographical areas, such as the Amazon basin the case of cocoa, the Congo basin in the case of oil palm, or the savannah zone of West Africa in the case of Shea butter nut. Morphologically taxa may show extraordinary morphological uniformity throughout such geographic ranges or they may be differentiated into geographical races, as in the case of peach palm and pummelo.

Other species have much more restricted distributions, such as rambutan in Malaysia and adjacent areas, and other may be even more localized or sparsely distributed with very low densities, such as some minor domesticated species related to the major citrus or mango fruit species.

There has been a tendency to try to store useful tropical woody species into the centers of diversity of cultivated plants, but this is not overly helpful. Since the species are mostly wild they occur particularly in those areas of high biological diversity, rather than high agricultural activity. The patterns of genetic variation are far less known than for arable crops. For instance, in the latter, few genepools contain species which defy clear definition, yet in tropical woody plants there are cases where this is true, due to reticulate morphological variation associated with ecotypic adaptation and specialization maintained by self-compatibilities (Leenhouts, 1968).

Whatever the spatial distribution of variation within a targeted tropical woody species, and in the absence of information on gene flow, effective population size and breeding systems, sampling for genetic conservation and use can only be based on pragmatic appreciation of geography and ecology. For longer term preservation of genepools, *in situ* conservation would be the major route to go, and this is obviously of importance for forest trees. However, species under some form of cultivation, where there are clear ecological and genetic links between the rainforests and back yard cultivation, provide additional opportunities for exploration and sampling, even when natural populations are being depleted due to loss of ecosystems. Nonetheless, there is a danger in over-reliance on primitively cultivated materials, because a degree of homogenization is likely to have occurred in the shorter lived species due to continual harvesting and reestablishment. An element of random sampling for genetic conservation and exploitation is clearly justified, but the majority of tropical tree crops have been sampled and used in a biased way (this will be discussed further below).

Scientific constraints are huge for this type of material. On the one hand, it is impossible to sample large numbers of populations across the regions of diversity, with many individuals in each sample. On the other hand, many cannot easily be conserved *ex situ*, and the problems of regeneration of collected samples often defy practicalities.

Genetic Variation of Populations

Most information on population genetics of trees has concentrated on measuring the level and distribution of genetic variability, isozyme loci in particular. Of course, such measures may not always parallel the amount of variability in the rest of the genome (Muona, 1990). Research is still scanty for tree crops, but studies on tropical trees in Panama, Costa Rica, Southeast Asia and Brazil by Drs. Hamrick, Bawa, Ashton and Kageyama, respectively, have shown that most species contain considerable within-population variation. Evidence from isozyme studies on variation within populations shows a much larger genetic diversity index for long-lived woody species compared to annuals and short-lived herbaceous perennials.

Even less is known about genetic differentiation between populations. However, in Brazil nut, there appeared to be little divergence among widely separated populations (Buckley *et al.*, 1988). In general, out-crossing species show higher variation within populations but lower variation among populations depending on pollinators (with animal pollination being intermediate with the opposite situation in self-pollinators) (Kageyama, 1990; Hamrick, 1983). And there is strong predominance of cross-pollination in the bulk of tropical tree species (Bawa *et al.*, 1985). The distribution of total diversity among populations shows annuals and short-lived herbaceous species to have the lowest genetic diversity within populations, but the highest amount of diversity among populations; woody perennials had values of the latter 3-4 times lower (Hamrick and Godt, 1990). This situation affects exploration, sampling and conservation. However, exploitation depends on knowledge gained in utilization of incompatibility mechanisms, such as in mango, durian, or carambola, or incidence of apomixis such as in langsat, some citrus or mangosteen. It also depends on the directional change in breeding systems which can be manipulated (Allard, 1975) or good knowledge of various breeding systems across the areas of diversity, as in the case of cocoa or rubber. The scientific community is only at the start of describing genetic structures of tropical tree crop populations, and virtually no information exists on release of variation over time in relation to reproductive output and seed dispersal and level of variation from different reproductive episodes (Bawa *et al.*, 1990).

Isozyme studies on samples of tree crops held in research collections have rarely been tied to patterns of diversity of wild or semi-wild populations. The studies are usually used to identify "cultivars" as in the case of mango, cherimoya or avocado; they may be part of biotechnological developments of crops such as papaya and *Citrus*; or they are used in taxonomy as in *Citrus*, banana or coffee.

Application of knowledge of patterns of diversity translate directly into the sampling strategies to be adopted when collecting germplasm. In many cases, the scarcity of resources remaining in the field and a degree of urgency to salvage samples for utilization require careful planning of field work. Over the past 20 years theoretical and practical considerations have led to the adoption of procedures which permit the acquisition of the maximum amount of practically useful variability in a particular species while keeping the numbers of samples practicable. For field crops the "aim" of the work is to include at least one copy of each variant in the target population with a frequency greater than 0.05 (Marshall and Brown, 1975). Numbers of individuals collected per site are therefore in the order of 50-100. Such numbers cannot be easily applied to forages nor to tree and other wood species.

Essentially the principle of random sampling versus biased sampling is necessary since genetic resources conservation collections require the locally common alleles rather than rare alleles. The former are alleles favored by natural or artificial selection; the latter may have conspicuous expression but are often not associated with factors favored in the target environment. Biased sampling is justified by users where the rare type may represent something special, such as a dwarf form of a normally large tree fruit or an ideotype which it has been decided needs to be developed. As well as random sampling, the sample is treated as a population sample, and grow-outs and resulting work for genetic conservation need to retain the population structure.

Clearly, application of principles such as random sampling and maintenance as populations pose problems for tropical perennials, due to the size of plants and their distributions. This is complicated by the following (Hawkes, 1980):

1) Most temperate and tropical fruit frees, many timber and nut trees, and economic trees and shrubs possess seeds which cannot be dried and held at low temperatures for storage (i.e., they are "recalcitrant"). Many are short-lived and require sowing very quickly.
2) In many cases hardwood is generally the most convenient material to collect.
3) Collecting strategy must be related to subsequent storage; the numbers of seeds proposed for field crops is far too large for tree crops.
4) Really reliable population sampling for species with recalcitrant seeds cannot be undertaken -- to apply the accepted sampling principles -- until storage systems such as *in vitro* genebanks are available.
5) In tropical forests, many species are sparsely scattered, and in many cases only one or two individuals can be found in every 10 hectares. Samples of these can only be done when the trees are found, and there can be no thought of population sampling at all.

As a result, some pragmatic rules of thumb can be developed. For wild materials seeds from up to 10-15 individuals in a specific area (up to 10 hectares) are collected (taking as many seeds as possible), and those are bulked into a single sample. For cultivated materials the site, such as a village, is treated as the population, and 10-15 individuals are bulked. If hardwood is collected, again this is from 10-15 individuals in the same 10 hectares, except for cultivars where evidence of clonal propagation is present; these are kept separate. Ten-hectare sites or villages are sampled depending on climatic, altitude or soil differences.

Collected materials have to be taken back to base and form an *ex situ* germplasm collection. Because of the nature of the materials, in some cases also quarantine restrictions, a great deal of genetic material tends to be lost. For that reason, I urged the International Board for Plant Genetic Resources (IBPGR) to initiate research on alternative collecting methods, and the past decade has shown remarkable advances. Traditional methods of handling cuttings in transit needed modification, since germplasm collecting in remote areas can be costly. One line of research has been to modify traditional methods for low cost alternatives using super-absorbent products and surface sterilizants (Aronson, Wickens and Birnbaum, 1990). Another line of research was to develop a simple, low cost *in vitro* holding of materials which, on arrival at base, could be propagated or transferred to *in vitro* systems for propagation and even *in vitro* storage. By the mid-1980s *in vitro* collecting was under test for cocoa, citrus, breadfruit, cassava, temperate fruits, African forage grasses which are shy-seeders, and others (Withers, 1987). One major advantage of the *in vitro* approach is that disease indexing and *in vitro* propagation can be brought together to provide a novel and safer quarantine system and avoid destruction of valuable genetic samples. Attention to germplasm collecting has indirectly resulted in acceleration of molecular indexing standards for disease agents and the development of clean-up procedures for a wide range of vegetatively propagated material.

SCIENTIFIC RESEARCH AREAS

Constraints to more rapid use of tropical tree crops relate to the needs for more focused research in three areas: understanding diversity, management of conserved populations, and exploiting diversity.

Diversity

In many cases, the use of traditional taxonomic methods has given fair ideas of the partitioning and distribution of diversity. A good example is the genus *Carica* (papayas) of which there are 21 species and at least 6 have been domesticated. The major one, *C. papaya*, probably had a hybrid origin involving 5 other species and was once thought to have originated in C. America. However, taxonomic survey indicates it almost certainly originated from an area extending along the eastern flanks of the Andes from Ecuador to Venezuela (Prance, 1984). Diversification of cultivars on other continents or movement through C. and S. America is evidenced from variation in primitive cultivars. Nonetheless, valuable as such studies are, we do not learn a great deal about genetic variation which should be conserved or utilized, except for interesting morphological variants. This is true of recent work on taxonomy of the mango genus and important secondary forest products such as bamboo and rattan, where taxonomy is providing little information on infraspecific variation.

As outlined above, techniques are now available to measure variation using biochemical and molecular techniques. Programs are needed which re-focus such screening away from examining *ex situ* collections, unless they are very comprehensive, to populations in particular regions. Since many of the regions of interest contain a range of species of interest, it would seem logical to consider rapid screening programs in a limited number of biological research centers which have the equipment and expertise. However, this needs to be carefully planned, because without information on the genetic correlations among individuals, the genetic basis of phenotypic variation has to be established through transplant gardens.

The genetic structure in plant populations results from interactions between mating systems, gene flow and selection. The screening would not unravel the role of each, except from studies over time by examining genetic relationships of individuals and parentages. In essence, what is needed are surveys: first, on a coarse grid basis, correlated with ecogeographical parameters and second, more intensive screening in specific areas. It requires standardization on parts of plants to be sampled, standardization of transport of samples and their analysis.

Many agricultural traits of tree crops, especially fruits, are due to multiple genes, and hence genetic studies are complex. There was an explosion of interest in variation as the biochemical techniques became available, because they allowed extra single-gene markers with which to study evolutionary processes: single-gene controlled morphological traits to use in quantitative estimates of gene flow or adaptive patterns in relation to ecotypic differentiation or other investigations. However, what are lacking are broad surveys of target species over wide areas and regions with measures of levels of variation within and between populations.

Some discussion has taken place on the potential of sampling species over wide areas and analyzing at the DNA level (IBPGR, 1988). More experimental work is needed when the interest is infraspecific diversity; differences in restriction fragments may be low in cultivated plants, e.g. tomato, or high, e.g. maize (Bernatsky and Tanksley, 1989). More work is needed on woody species and the results used in assessing how much emphasis should be on isozyme analysis and how much on the use of DNA markers.

Recent research has shown that it is now indeed possible to survey areas on the basis of site characteristics and gather herbarium-type specimens, analyze these in conjunction with each other (the latter using restriction enzyme digestible DNA), then plan the most efficient field collecting. Work on the genus *Musa* in Papua New Guinea, an area of diversity, has used RFLP analysis in this way to understand variability in an array of parthenocarpic edible forms collectively known as *M. fehi*, and their genetic relationships with a range of wild species, including confirmation of a progenitor species (Gawel, Jarret and Whittemore, 1991; Jarret, Gawel, Whitmore and Sharrock, unpublished).

Management of Conserved Populations

For many tropical trees and tree or woody crops, conservation for future genetic use is going to rely heavily on conservation of target populations in ecosystems, i.e. *in situ*. However, much of the design of reserve areas relates to maximum species numbers rather than target populations of one component species. The fact that a useful species is present does not mean that the genetic diversity is necessarily of high utility. Additionally, maintenance of the target species might well require specific management and monitoring of population fitness over time. Without the latter there will inevitably be *ad hoc* and/or biased sampling for future genetic conservation and utilization purposes. In addition, microenvironmental selection, which creates sharp patterns of genetic variation in populations (Allard *et al.*, 1972; Bradshaw, 1984), must not be confused with the broader patterns of ecogeographic differentiation. Current interest in spatial autocorrelation on data from single populations (Epperson, 1990) should not detract from the need to devise procedures for monitoring overall genetic variation, as total variance, across a range of populations.

Wise management of natural populations has to go hand in hand with the close integration of genetic resources research and nature conservation programs. Special genetic reserves are necessary to cover certain useful species not adequately covered in conservation of ecosystem programs. Nonetheless, these special concerns will become increasingly difficult to justify in terms of maintenance, and perhaps only those managed to benefit local people will be viable. The strategies should emphasize species which cannot be conserved *ex situ* due to recalcitrant storage behavior of seeds, and a whole range of methods needs to be devised to utilize extractive reserves for better conservation (Williams, 1991).

Enough knowledge exists to develop procedures and protocols to manage conserved populations, but interdisciplinary groups are needed to establish guidelines based on the best science.

Colonizing Species

A great number of species of genepools of tropical tree crops exhibit a degree of weedy tendency, which means that they are adapted to a wide range of ecologies and often become "camp followers" in relation to habitations. This must have been one reason for their domestication and over time many tree crops have established and spread in habitats and regions not previously occupied. Information on the genetic consequences of colonization is useful in understanding past domestications processes, but more importantly, for future ecological and evolutionary potentials which can be manipulated through purposive breeding.

In many cases, small populations resulting from founder effects and bottlenecks exhibit particular patterns of variation, and these are often the primitively cultivated forms exhibiting genetic drift. Many of these populations have been sampled in a non-random way to build up the *ex situ* collections used by breeders. Many of these collections are expensive to maintain and in some cases, it would be better to start sampling from scratch.

Although there is a respectable knowledge based on the genetics of colonizing plants, much of it is based on herbaceous species. Due to tree crops being mostly little differentiated from the wild genepool, great advantages would accrue from experimental work in relation to rapid artificial domestication and selection. Assumptions that central habitats are environmentally more benign and marginal sites colonized result in genetically differentiated and less variable populations due to strong selection or genetic drift (Antonovics, 1976; Barrett and Husband, 1990) (correlated with the ecogeographic background), need testing for tropical woody species.

Woody species are not the easiest material to work with, but many, such as papaya, guava, *Leucaena* and others are fast growing, and measuring genetic structures of populations in space and time should not deter the need for this focused research.

One scientific issue could be resolved fairly readily, with great practical application to germplasm sampling strategy and the build-up of *ex situ* collections containing representative patterns of variation. This issue relates to a categorization of tree crop species regenerative behavior in natural habitats. For instance, of primary tree species, those which are shade tolerant germinate in the closed understorey and exist as suppressed juveniles in clumps, because dispersal is limited. These have clumped distributions, and there are wide variations in seed dormancy (Garwood, 1983; Liew and Wong, 1973). Any tree crop from this category could be stored in a cost-effective way in an innovative genebank (Hawkes, 1981). However, many of the useful plants are shade intolerant, and juveniles are found in gaps which are colonized by seed and are quick to invade large natural gaps and areas cleared by humans. Such species tend to have stronger dormancy patterns than primary species, and soil seed banks are common (Brokaw, 1985). The third type includes understorey plants such as many of the palms (Whitmore, 1978).

Patterns of gap regeneration over space and genetic differentiation related to disturbance as a source of spatial heterogeneity might be more important than niche partitioning of diversity. Almost certainly, width of environmental tolerances of many tree crops correlates with the types of natural patterns of regeneration. Although there is a respectable body of literature on natural disturbance and patch dynamics in tropical forests, there is less for dry-season deciduous forests. Few correlations have been sought between such spatial heterogeneities and the adoption of primitive farmers of certain woody species which, due to some inherent capacity, could become part of the farming systems associated with, for example, slash and burn agriculture. Most of such species are pioneers, relatively shade intolerant and abundant seeders. Almost certainly, many tree crops were developed from genepools in spatially subdivided and temporally unstable environments. Whether patterns of variation evolved through selfing and the rapid fixation of successful genotypes or through higher rates of outcrossing, more information on genetic versus phenotypically adaptive strategies would be valuable to determining what exactly to collect and conserve as genetic resources.

GENETIC RESOURCES NEEDS

For reasons outlined above, there is a need for the accelerated collection of species specific plant germplasm of tropical tree crops to enable *ex situ* conservation and more rapid enhancement of germplasm. To do this properly, there must be increased scientific focus on related biological inventories and ecogeographical surveys in the areas of diversity. In some instances and with some species these activities could conceivably be enacted as part and parcel of accelerated field projects. Such projects will interact closely with wider conservation efforts, since they involve multiple land use and conservation.

Plans have to be developed on a crop-by-crop basis in order to target the best possible capture of variability and its securest possible conservation. Each step requires the adoption of complementary methods, some for which the technology exits, others where further research and development are necessary.

Apart from some important crops such as rubber, cocoa, banana, the germplasm collections are woefully inadequate for most tropical tree crops. In some cases they hardly exist, neither *ex situ* nor carefully planned *in situ*. Whole industries are unsustained by genetic resources conservation despite millions of people being dependent. In each case scientific work is needed to answer the questions what and how to sample and how to conserve.

Since the accelerated scientific work will be strategic research, use of the newer technologies will streamline utilization. For instance, *in vitro* conservation of segments of diversity is closely linked to propagation and furthermore *in vitro* systems lend themselves to new opportunities for characterization, evaluation, wide-crossing and selective gene transfer.

Almost certainly forest tree breeding will require better organized systems for broad scale provenance testing of species of interest, but there is no global strategy for low cost, rapid screening of hundreds of species (NRC, 1991). With tree crops there are vested interests for rural development, sustainable use of the environment, conservation, producers and consumers. Planning and implementing re-focused genetic and other scientific research in a strategic way is an urgent challenge.

Of the genetic resources work ongoing at present and of the international activities of the past 2-3 decades, no group of plants has been so neglected as tropical woody perennials. In part this is understandable because breeding efforts are low for many, and materials are difficult to conserve. I doubt greatly if activities will be enhanced following the well tried and tested method of scientific research tied to public sector support, which has proved successful for food crops. The input of the private sector, following examples of consumer organizations for cocoa and products for others will be necessary, and for the array of minor species of interest to community well being and local sustainable practices, local and regional nongovernmental organizations might well take the lead.

REFERENCES

Allard, R. W., 1975, The mating system and microevolution, *Genetics*, 79:115-126.

Allard, R. W., Kahler, A. L., and Weir, B. S., 1972, The effect of selection on esterase allozymes in a barley population, *Genetics*, 73:489-503.

Antonovics, J., 1976, The nature of limits to natural selection, *Ann. Miss. Bot. Gard.*, 63:224-247.

Aronson, J., Wickens, G., and Birnbaum, E., 1990, An experimental technique for long-distance transport of evergreen or deciduous cuttings under tropical conditions, *Pl. Genetic Resources Newsl.*, 81/82:47-48.

Barrett, S.C.H., and Husband, B. C., 1990, The genetics of plant migration and colonization, *in*: "Plant Population Genetics, Breeding, and Genetic Resources," A.H.D. Brown, M. T. Clegg, A. L. Kahler, and B. X. Weir, eds., Sinauer, Sunderland, MA.

Bawa, K. S., 1976, Breeding of tropical forest hardwoods, *in*: "Tropical Trees: Variation, Breeding and Conservation," J. Burley and B. T. Styles, eds., Academic Press, London.

Bawa, K. S., Ashton, P. S., and Nor, Salleh M., 1990, Reproductive ecology of tropical forest plants: management issues, *in*: "Reproductive Ecology of Tropical Forest Plants," K. S. Bawa and M. Hadley, eds., UNESCO/Parthenon, Paris; Carnforth, UK.

Bawa, K. S., Perry D. R., and Beach, J. H., 1985, Reproductive biology of tropical lowland rain forest trees. 1. Sexual systems and incompatibility mechanisms, *Amer. J. Bot.*, 72:331-345.

Bernatsky, R., and Tanksley, S. D., 1989, Restriction fragments on molecular markers for germplasm evaluation and utilization, *in*: "The Use of Plant Genetic Resources," A.H.D. Brown, O. H. Frankel, D. R. Marshall, and J. T. Williams, eds., Cambridge University Press, Cambridge.

Bradshaw, A. D., 1984, Ecological significance of genetic variation between populations, *in*: "Perspectives on Plant Population Ecology," R. Dirzo and J. Sarukhan, eds., Sinauer, Sunderland, MA.

Brokaw, N.V.L., 1985, Treefalls, regrowth and community structure in tropical forests, *in*: "The Ecology of Natural Disturbance and Patch Dynamics," S.T.A. Pickett and P. S. White, eds., Academic Press, San Diego.

Garwood, N. C., 1983, Seed germination in a seasonal tropical forest i Panam: a community study, *Ecol. Monogr.*, 53:159-181.

Gawel, N., Jarret, R. L., and Whittemore, A., 1991, Restriction fragment length polymorphisms (RFLP)-based phylogenetic analysis of *Musa*, *Syst. Bot.* (in press).

Hamrick, J. L., 1983, The distribution of genetic variation within and among natural forest populations, *in*: "Genetics and Conservation," C. M. Schonewald-Cos, S. M. Chambers, B. MacBryde, and W. L. Thomas, eds., Cummings, Menlo Park, CA.

Hamrick, J. L., and Godt, M.J.W., 1990, Allozyme diversity in plant species, *in*: "Plant Population Genetics, Breeding, and Genetic Resources," A.H.D. Brown, M. T. Clegg, A. L. Kahler, and B. S. Weir, eds., Sinauer Associates, Sunderland, MA.

Hawkes, J. G., 1980, "Crop Genetic Resources Field Collection Manual", IBPGR/Eucarpia, Wageningen.

Hawkes, J. G., 1981, Genetic conservation of "recalcitrant species" -- an overview, *in*: "Crop Genetic Resources: The Conservation of Difficult Material," L. A. Withers and J. T. Williams, eds., IUBS (Série B42), Paris.

IBPGR, 1988, *Report of the Program Committee*, Rome Italy.

Kageyama, P. T., 1990, Genetic structure of tropical tree species in Brazil, *in*: Reproductive Ecology of Tropical Forest Plants," K. S. Bawa and M. Hadley, eds., UNESCO/Parthenon, Paris; Carnforth, UK.

Leenhouts, P. F., 1968, A conspectus of the genus *Allophylus* (Sapindaceae). A problem of the complex species, *Blumea*, 15:301-358.

Liew, T. C., and Wong, F. O., 1973, Density, recruitment, mortality and growth of dipterocarp seedlings in virgin and logged-over forested in Sabah, *Malays. For.*, 36:3-15.

Marshall, D. R., and Brown, A.H.D., 1975, Optimum sampling strategies in genetic conservation, *in*: "Crop Genetic Resources for Today and

Tomorrow," O. H. Frankel and J. G. Hawkes, eds., Cambridge University Press, Cambridge.

Muona, O., 1990, Population genetics in forest tree improvement, *in*: "Plant Population Genetics, Breeding, and Genetic Resources," A.H.D. Brown, M. T. Clegg, A. L. Kahler, and B. S. Weir, eds., Sinauer, Sunderland, MA.

National Research Council, 1991, "Managing Global Genetic Resources: Forest Trees," National Academy Press, Washington, DC.

Smith, N.J.H., Williams, J. T., Plucknett, D. L., and Talbot, J. P., "Tropical Forests and Crop Genetic Resources," (in press), Cornell University Press, Ithaca, NY.

Whitmore, T. C., 1978, Gaps in the forest canopy, *in*: "Tropical Trees as Living Systems," P. B. Tomlinson and M. H. Zimmermann, eds., Cambridge University Press, London.

Wilken, G. C., 1991, Sustainable agriculture is the solution, but what is the problem? Occasional paper no. 14, BIFADEC/AID, Washington, DC.

Williams, J. T., 1991, Genetic resource conservation in natural habitats: biological and socioeconmic dimensions, *in*: "Strategic Approaches to Conserving Genetic Resources in Natural Habitats," J. I. Cohen and C. S. Potter, eds., AAAS, in final preparation, Washington, DC.

Withers, L. A., 1987, *In vitro* methods for germplasm collecting in the field, *Pl. Genetic Resources Newsl.*, 69:2-6.

Wood, P. J., 1976, Exploitation of tropical plantations and the need for seed and genetic conservation, *in*: "Tropical Trees: Variation, Breeding and Conservation," J. Burley and B. T. Styles, eds., Academic Press, London.

THE NARANJILLA (*SOLANUM QUITOENSE*), THE COCONA (*SOLANUM SESSILIFLORUM*) AND THEIR HYBRID

Charles B. Heiser

Department of Biology
Indiana University
Bloomington, Indiana 47405

INTRODUCTION

The Solanaceae has been one of the more important families in providing useful plants for humankind. By far the greatest number of these comes from tropical America, white potatoes, tomatoes and chili peppers being the best known. In recent years the fruits of two others, the tree tomato *(Cyphomandra betacea)* and the pepino *(Solanum muricatum)* have begun to appear in our markets, both imported from New Zealand where they had been introduced from tropical America. To my thinking two other members of the family, the naranjilla (*S. quitoense*) and the cocona (*S. sessiliflorum*), are equal or superior to the last two, but they have yet to reach markets outside of Latin America. These species, both placed in the section *Lasiocarpa* of *Solanum,* are the subjects of this paper.

The naranjilla, or *lulo* as it is often called in Colombia, is a shrubby perennial, nearly 2 m tall, with large purple veined leaves and white corollas. The rotund berries are covered with short, stiff hairs which are fairly readily deciduous so that they are glabrous or nearly so when they reach the market. At maturity the fruits, 4.5 to 5.5 cm in diameter, are orange, which is responsible for the Spanish name meaning "little orange." The flesh or pulp of the fruit, however, is green (Whalen, Costich and Heiser, 1981). The juice is used most commonly for drinks. The fragrance and flavor are unique, but some people have claimed a resemblance to a mixture of pineapple and strawberries, others to that of citrus fruits. The fruit is rich in vitamins A and C (Romero-Castañeda, 1961).

In the sixteenth century Cobo (Patiño, 1962) described plants from the regions of Popayan and Quito that are thought to represent the naranjilla. To this day Colombia and Ecuador, where the plants are generally grown at altitudes of 1000 to 1500 m, remain the principal producers. It is also grown in

Venezuela and at one time in Peru. Less than a half century ago it was introduced to Panama and Costa Rica, where it is now well established, and more recently to Guatemala (Heiser, 1985). Attempts to establish it as a crop in Florida were unsuccessful, although the early trials appeared promising (Ledin, 1952). The plant was promoted in a recent publication (National Research Council, 1975) and since that time I have had requests for seeds from people in tropical Asia and Africa.

For a domesticated plant, the naranjilla shows extremely little variability (Heiser, 1972; Whalen et al., 1983). Plants grown in Ecuador are unarmed whereas to the north of that country, they generally have small prickles on the leaves and stems (Schultes and Romero-Castañeda, 1962). In Ecuador five "varieties" are recognized: *agria, Baeza, Baeza roja, bola,* and *dulce,* distinguished by very slight differences in the fruits. Because little breeding work has been done with the naranjilla, the plants grown today probably differ little from those found by the Spanish when they arrived in the Americas. Although the naranjilla occasionally escapes from cultivation to establish weedy populations, no truly wild type is known.

The cocona, also known as *topiro* or *tupiru* and in Brazil as *cubiu,* became known to Europeans some half a century later than the naranjilla (Patiño, 1962) and today is widely cultivated in the upper Amazon basin at altitudes up to 700 m or slightly higher. It is a shorter plant than the naranjilla, and it has even larger leaves, generally with little or no purple pigmentation, and greenish-white corollas. The hairs on the berry are softer and more readily deciduous than those of the naranjilla. The mature fruits are orange to maroon and rather variable in shape and size from 4 to 9 cm in diameter. The fruit flesh is nearly white to pale yellow. The aroma and taste of the juice are less pronounced than that of the naranjilla. In addition to being used for juice, the fruits are also used as a vegetable. The fruit is rich in iron and contains vitamins A, C and niacin (Salick, 1989).

A wild type of the cocona (*S. sessiliflorum* var. *georgicum*), known from a few collections in Colombia, Ecuador and Peru, differs from the domesticate by being prickly and having much smaller fruits. In Ecuador, a weedy or wild unarmed type, has fruits about 3 cm in diameter, and is known as *naranjilla jíbara.* Clement (1989) considers the cocona one of the several fruits domesticated in western Amazonia.

My involvement with these species began in 1962. On my early trips to South America I found the naranjilla very abundant and inexpensive in markets in Ecuador but over the years it became less numerous and more expensive. The principal reason for this apparently is the large number of diseases and pests to which the naranjilla is subject. One of the most serious of these is root knot nematodes. Resistance to root knot nematode was discovered in *S. hirtum* (Heiser, 1971), a species that crosses readily with *S. quitoense;* and although some effort has been made to incorporate this trait in the naranjilla, resistant varieties have yet to be grown commercially unless it is in Colombia, a country from which I have no recent information.

In 1982 a conference on the naranjilla was held in Quito (Anon., 1984). Before the conference Saúl Camacho took me to the experiment station for the naranjilla at Chaco and there for the first time I saw hybrids of the naranjilla and the cocona. Although more like the cocona in general aspect, the hybrids had a number of intermediate characters. The hybrid was of great interest to me, for I had tried to make it for many years without success. Even more surprising was the account of its origin, for it had been made by a campesino, Raúl Viteri. He had heard of plant hybridization being used to improve plants so he had crossed the two species, using *naranjilla jíbara* as the pollen donor. Months later under the naranjilla he found seedlings whose leaf coloration was intermediate between that of the putative parents. He transplanted some of the seedlings and found that the fruits were not attacked by diseases or insects. The fruits although smaller than naranjillas had a similar color and taste. The seeds were not viable but vegetative propagation was successful (Torre and Camacho, 1981). I made a brief visit to Ecuador in 1985. I didn't visit markets in Quito but in those at Ambato the only fruits I saw were from the hybrid which were called naranjilla or *naranjilla chica* (small). They were readily recognizable as the hybrid for they were only half the size of true naranjillas and contained very few or no seeds. On this trip I also learned that the experiment station at Chaco had been abandoned by the government because of a recent financial crisis.

On my return to Indiana I renewed my efforts to produce the hybrid and numerous attempts during the next five years were again unsuccessful. No fruit ever developed on the naranjilla as a result of the cross, and although the reciprocal cross yielded several fruits, not a single seed was found.

My next trip to Ecuador was in 1988, and I did see a number of what appeared to be normal naranjillas. On two occasions our hostess in Quito purchased some and I was puzzled because I failed to find a single seed in any of the fruits. The answer was provided when Jorge Soria and I made a trip to the new station that had been established for the naranjilla at Palora. Although Palora at an altitude of 950 m is too low for the best growth of the naranjilla, the hybrid thrives there. We learned that a few years earlier some weeds near a field where the hybrids had been growing had been sprayed with 2-4,D. It was later noted that the hybrids downwind of the spray had fruits about twice the size of the other plants. Since that time hybrids in flower have deliberately been sprayed with a very dilute solution of 2-4,D to produce larger fruits, such as those we had seen in Quito. In 1990 it was estimated that two thirds or more of the crop in Ecuador resulted from hybrids sprayed with 2-4,D (J. Soria, pers. commun.).

Again following my return to Indiana I attempted to make the hybrid, using some new accessions that I had acquired from Víctor Rodríguez as well as my older ones. In addition to my own desire to obtain the hybrid, there were other reasons for producing it as well. First, the hybrid being grown in Ecuador was very uniform, possibly constituting a single genotype, and hence likely had great genetic vulnerability. Secondly, I thought that it might be possible to produce a larger fruited hybrid by using a large fruited cocona as a parent eliminating the necessity of spraying with 2-4,D. Again a number of fruits was formed when the cocona was used as the female parent. In order to attempt embryo rescue, my colleague Martha Crouch examined a very young fruit in search of embryos. None was found. The other fruits were allowed to reach maturity and in one of them two partially developed seeds were found. These were planted on a nutrient medium (Miller, 1969) supplied by Carlos Miller. Both germinated and were transplanted to soil six weeks later. The parents of this hybrid were a domesticated cocona which I had collected at Yanzatza, Ecuador (8255) and the variety *Baeza roja* of the naranjilla.

The hybrids (8919) grown in the greenhouse were more vigorous than either parent and morphologically similar to Viteri's hybrid except in two respects. The fruits were much larger, slightly larger than naranjillas, and the fruit flesh was orange instead of green. The increase in size was expected, for the cocona parent had fruits over 7 cm in diameter, but the orange color was not anticipated. Little is known of the genetics of this group of *Solanum, so* the reason for the different fruit colors remains speculative. One explanation may involve the difference in the parentage of the two hybrids. Viteri used a wild cocona with small fruits that externally are yellow-orange whereas I used a domesticate with maroon fruits; both, however, have a very light colored flesh. The variety of the naranjilla used by Viteri is unknown, but considering the region where he made the hybrid I would judge that it was *dulce.* The fruit of that variety has a deep green flesh, whereas *Baeza roja,* the male parent of my hybrid, has a fruit with much lighter green flesh. Another possible explanation in view of the different female parentage of the two hybrids is that a maternal effect may be operating. Viteri's hybrid has a light green fruit flesh approaching that of the naranjilla. Mine is nearer to that of the cocona, but not identical to it by any means. Salick (1989) has postulated a maternal effect to account for the size and shape of fruits in hybrids involving varieties of cocona with different fruit types.

My hybrids gave pollen stainabilities of 28 and 38%. No seed was set on selfing, in sib crosses or in backcrosses made to either parent in both directions. A few seeds, however, were found in some of the fruits in a greenhouse open to pollinators in which several accessions of both of the parental species as well as other species of the section were growing. The low fertility of the hybrids can probably be explained as the result of meiotic irregularities. At diakinesis in one hybrid the following pairing configurations were observed: 12_{II} (3 cells), $11_{II}2_{I}$ (3), $10_{II}1_{III}1_{II}$ (8), $10_{II}1_{IV}$ (13), $10_{II}4_{I}$ (7), $9_{II}1_{III},3_{I}$ (7), $9_{II}1_{IV}2_{I}$ (6) and $9_{II}6_{I}$ (1).

The aroma and the flavor of the fruit of the new hybrid are excellent. They are very similar to that of the naranjilla but not identical to it. This opinion is based on my own evaluation and that of three others. I do not yet have reports on it from the people who are growing it in Florida and Ecuador. Possibly the different color of the juice may hinder its acceptance in areas where people expect it to be green; on the other hand, it may make it more acceptable in new areas. Although I hesitate to make predictions, I feel that the juice of both the naranjilla and the hybrid could become popular in the United States and Europe. As for the cocona, it may become more widely grown in the tropics, but I do not believe that either as a juice or as a vegetable, it can compete with the many other juices and vegetables now available in the northern temperate zone.

The enthusiasm for growing the hybrid in Ecuador could mean that attempts to improve the naranjilla by crossing will be abandoned. I hope that this does not prove true. Although there is very little variability in this species, it is now known to hybridize with five other species (Heiser, 1989) so that genes for its improvement may be available. Through backcrossing it may be possible to transfer desirable genes into the domesticated species. Such a procedure can readily be accomplished by scientists in the Third World who may lack the facilities to pursue newer more sophisticated methods for gene transfer.

REFERENCES

Anonymous, 1984, "Memorias de la primera conferencia internacional de naranjilla," Instituto Nacional de Investigaciones Agropecuarias, Quito, Ecuador.

Clement, C., 1989, A center of crop diversity in western Amazonia, *BioSciences* 39:624-630.

Heiser, C., 1971, Notes on some species of *Solanum* (sect. *Leptostemonum*) in Latin America, *Baileya* 18:59-65.

Heiser, C., 1972, The relationship of the naranjilla, *Solanum quitoense*, *Biotropica* 4:77-84.

Heiser, C., 1985, "Of Plants and People," University of Oklahoma Press, Norman.

Heiser, C., 1989, Artificial hybrids in *Solanum* sect. *Lasiocarpa*, *Syst. Bot.* 14:3-6.

Ledin, R., 1952, The naranjilla (*Solanum quitoense* Lam.), *Proc. Florida State Hort. Soc.* 65:187-190.

Miller, C., 1969, Cytokinin production by mycorrhizal fungi, in: "Proceedings of the first North American Conference on Mycorrhizae," Misc. Publ. 1189, U.S. Dept. Agric., Washington, D.C.

National Research Council, 1975, "Underexploited tropical plants with promising economic value," National Academy of Sciences, Washington, D.C.

Patiño, V., 1962, Edible fruits of *Solanum* in South American historic and geographic references, *Bot. Mus. Leaf., Harvard Univ.* 19:215-234.

Romero-Castañeda, R., 1961, El lulo: una fructa de importancia economica, Agricultural *Tropical* 17:214-218.

Salick, J., 1989, Cocona (*Solanum sessiliflorum*) production and breeding potentials of the peach-tomato, in: "New Crops for Food and Industry," Wickens, G. and Day, P., eds., Chapman and Hall, London.

Schultes, R. and Romero-Castañeda, R., 1962, Edible fruits of *Solanum* in Colombia, *Bot. Mus. Leaf., Harvard Univ.* 19:235-286.

Torre, R. and Camacho, S., 1981, Campesino fitomejorador de naranjilla, *Carta de Frutales* no. 14, Instituto Nacional de Investigaciones Agropecuarias, Quito, Ecuador.

Whalen, M. and Caruso, E., 1983, Phylogeny of *Solanum* sect. *Lasiocarpa* (Solanaceae): Confluence of morphological and molecular data, *Syst. Bot.* 8:369-380.

Whalen, M., Costich, D. and Heiser, C., 1981, Taxonomy of *Solanum* section *Lasiocarpa*, *GentesHerb.* 12:41-129.

THE CONSERVATION AND UTILIZATION OF GENES FROM ENDANGERED AND EXTINCT PLANTS: DNA BANK-NET

Robert P. Adams

Plant Biotechnology Center,
Baylor University,
BU Box 7372, Waco, TX 76798

SUMMARY

The recent advent of more efficient methods for DNA extraction and amplification (via PCR) now make it both prudent and urgent that we begin to amass significant genomic DNA samples from endangered plants with the greatest priority given to tropical rainforest species. DNA Bank-Net is an association of institutions that have begun to accession DNA and DNA-rich materials for subsequent disbursement of genes and oligonucleotides. The formation and functions of DNA BankNet are presented.

INTRODUCTION

Since the first plant to plant gene transfer occurred in 1983 (Murai et al., 1983), genes have been transferred to plants from viruses (Nelson et al., 1988), bacteria (Barton et al., 1987; Della-Cioppa et al., 1987; Fischhoff et al., 1987), and even from mammals to plants (Lefebvre et al., 1987; Maiti et al., 1988). Genetic transfers are being performed in order to attain insect, bacteria, viral and fungal resistance, a more nutritionally balanced protein, more efficient photosynthesis, nitrogen fixation, and salt and heavy metal tolerance, to name a few. These kinds of gene transfers from one unrelated organism to another indicate that we must now view the world's genetic resources (genes, DNA) from a horizontal perspective in which gene transfers will cut across species, genera and family boundaries.

For example, a strain of cowpea, *Vigna unquiculata* (L.) Walp., discovered in a market in Ilorin, Nigeria, contains a protein that inhibits trypsin digestion by insects (Redden et al., 1984). This gene has been moved to tobacco (*Nicotiana*) where the trypsin- inhibiting gene is expressed and offers tobacco the same resistance against insects as in cowpea (Newmark, 1987). It is interesting to note that although a very active form of the gene has been found in a Nigerian cowpea, scarcely 100 of the world's 13,000 legume species have been examined for this gene. Yet, the tropical legumes, one of the most promising groups for the evolution of natural insecticides, will certainly be subject to considerable germplasm loss in the next decade.

The number of novel insecticides, biocides, medicines, etc. that could exist in nature is innumerable. Yet, the principal areas of diversity among plants, the lowland tropical forests, will have been cut or severely damaged within the next 20 years (Raven, 1987, 1988). The Amazon River system, for example, contains eight times as many species as the Mississippi River system (Shulman, 1986). Raven (1987, 1988) estimated that as many as 1.2 million species would become extinct in the next twenty years. The loss of plant species will mean a loss of potential plant derived pharmaceuticals, now estimated at $2 billion/year in the United States alone (U. S. Congress, 1987).

It also means a loss of genetic diversity present in and available to our current and potential crop species. Cultivated crops are extremely inbred for factors such as yield, uniform flowering and height, and cosmetic features of the products. This narrow genetic base has resulted in several disastrous crop failures. Ireland's potato (*Solanum tuberosum* L.) famine of 1846, which resulted in famine and the emigration of a quarter of its population, was due to the fact (Plucknett et al., 1987) that their potatoes had no resistance to the late blight fungus (*Phytophthora infestans*). This can be traced to the lack of genetic diversity in Irish potatoes, which had been multiplied using clonal materials from just two separate South American introductions to Spain in 1570 and to England in 1590 (Hawkes, 1979).

A more recent example is the southern corn leaf blight fungus (*Helminthosporium maydis*) in 1970 in the United States. Because almost all of the corn (*Zea mays* L.) in the United States was of hybrid origin and contained the Texas cytoplasmic male sterile line, our fields of corn presented an unlimited extremely narrow gene base habitat for the fungus. By the late summer, 1970, plant breeders were scouring corn germplasm collections in Argentina, Hungary, Yugoslavia and the United States for resistant sources (Plucknett et al., 1987). Nurseries and seed fields were used in Hawaii, Florida, the Caribbean, and Central and South America to incorporate the resistance into hybrid corn in time for planting in the spring of 1971 (Ullstrup, 1972). Without these genetic resources this technological feat would not have been possible.

The National Cancer Institute (NCI) is now spending $8 million over the next 5 years for a massive plant collecting effort in the tropics to find anti-cancer and anti-AIDS virus compounds (Booth, 1987). The plant collectors will gather leaves and/or bark and air-dry the material for shipment to Maryland where it will be extracted and assayed against 100 cancer cell lines and the

AIDS virus. Yet, no genetic resources will be collected! When a promising compound is found, the plants will have to be recollected. For extensive testing (as well as commercial utilization), plantations will have to be established in the tropics to provide material.

Collections of plant specimens have been utilized for the formulation of our understanding of morphological variation among taxa. Indeed, without the great herbaria of the world, our knowledge of plant evolution would be fragmented at the least. As we have moved into the era of utilizing chemical data for systematic and evolutionary studies, methods of preserving plant materials for future (chemical) work have been largely ignored. We are usually content to file a voucher specimen to document our chemical studies. With the present level of support for plant collections, it is unlikely that much of the world's plant species can be preserved by freezing so that scientists might have access to the study of secondary compounds, enzymes, or DNA/RNA in the coming centuries.

FORMATION OF DNA BANK-NET

Concurrent with the advancements in gene cloning and transfer, has been the development of technology for the removal and analyses of DNA. DNAs from the nucleus, mitochondrion, and chloroplast are now routinely extracted and immobilized onto nitro-cellulose sheets where the DNA can be probed with numerous cloned genes. Recent advances in the technology for the extraction and immobilization of DNA, coupled with the prospect of the loss of significant plant genetic resources throughout the world, has led to the establishment of DNA Bank-Net, an international network of DNA repositories for the storage of genomic DNA on every continent.

A group of 18 scientists held the organizational meeting at the Royal Botanic Gardens, Kew, London, April, 1991 to share country and institutional experiences using *in vitro* biotechnology and particularly cryostorage of DNA and DNA-rich materials (Adams and Adams, 1991). Relatively few scientists were interested in a 'genetic insurance policy' when the idea of banking genomic DNA from plants was first proposed (Adams, 1988, 1990). However, currently there are 40 institutions (representing 25 nations and every continent, see Appendix I) that have expressed interest in DNA Bank-Net (Figure 1).

The conserved DNA will have numerous uses: molecular phylogenetics and systematics of extant and extinct taxa; production of previously characterized secondary compounds in trans-genic cell cultures; production of trans-genic plants using genes from gene families; *In vitro* expression and study of enzyme structure and function; and genomic probes for research laboratories.

STRUCTURE AND OPERATION OF DNA BANK-NET

At the organizational meeting of DNA Bank-Net, a task force was convened to define the functions of working (DNA dispensing) and reserve (base) nodes in the DNA Bank network. The group recommended the following functions (Adams and Adams, 1991):

Working (DNA dispensing) nodes:

a. Collection of plant material by taxonomists. This may be the primary function of a particular node or be in association with other organizations such as universities, botanic gardens, etc.
b. DNA extraction by molecular biologists or trained staff.
c. Long-term preservation of DNA-rich materials and/or extracted DNA in liquid nitrogen.
d. DNA analysis/gene replication by molecular biologists or trained staff.
e. Distribution of DNA (genes, gene segments, oligonucleotides, etc.).

Reserve (base) nodes:

a. Long term DNA preservation in liquid nitrogen and monitoring of potential DNA degradation.
b. Act as genetic reserve buffer for working nodes.
c. Replenishment of DNA if a working node experiences the catastrophic loss of storage parameters and DNA.

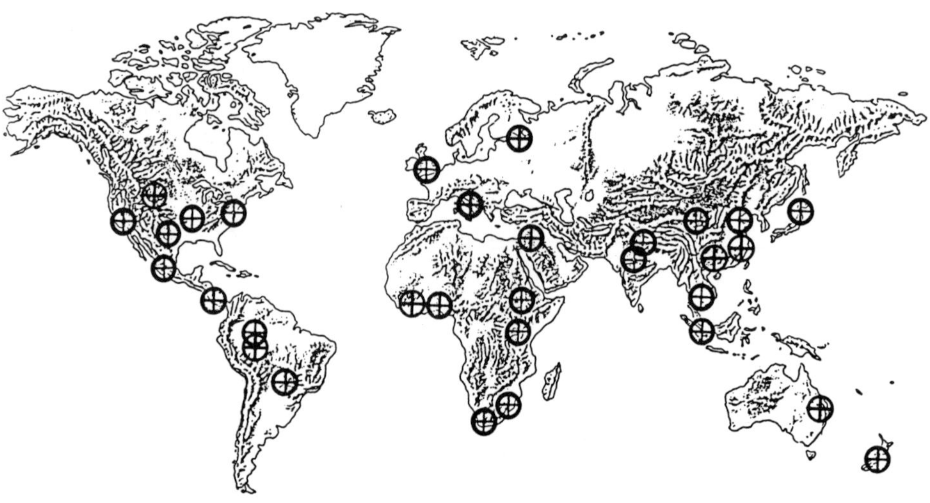

Fig. 1. Map of individuals/institutions currently interested in DNA Bank-Net.

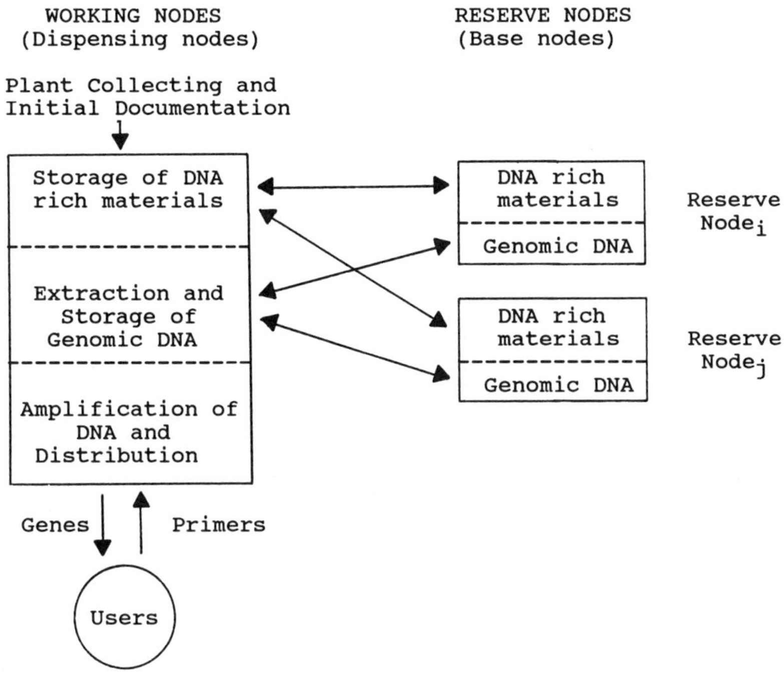

Fig. 2. Schematic representation of the flow of materials and the relationship between working (DNA dispensing), reserve (base) nodes and users.

Figure 2 depicts the relationship between working and reserve nodes. Note the projected flow of plant materials and DNA through the working (DNA dispensing) node. It is likely that some of the working nodes would be actively acquiring and/or dispensing DNA from some geographic area (ex. Africa), yet maintain separate cryovats, functioning as a reserve (base) node for another area (ex. South America).

GENERAL REQUIREMENTS FOR NODES IN THE DNA BANK-NET

The task group recommended (Adams and Adams, 1991) that the following were the minimum requirements for nodes:

Working (DNA dispensing) nodes:

Personnel: Taxonomists/collectors, biochemists/molecular biologists, technicians for practical work, capable administration

Equipment: Storage facilities (liquid nitrogen, cryovats), extraction facilities (centrifuges, gel electrophoresis, UV spectrophotometer, etc.), DNA Analyses and PCR duplication (PCR thermal cycler, micro-centrifuges, etc.), distribution system (packaging and mailing supplies), computer (database for inventory and correspondence).

Reserve (base) nodes:

Personnel: Technicians, capable administration.

Equipment: Storage facilities (liquid nitrogen, cryovats), computer (data base for inventory and correspondence).

Each DNA collection should be split initially into at least 2 or 3 portions. One sample (DNA-rich material or extracted DNA) should be stored at a working (DNA dispensing) node and another portion(s) be stored in at least 1 (one), but desirably 2 (two) back-up reserve (base) nodes. The reserve nodes should be in different countries and if possible on different continents to safeguard the DNA samples against various natural and man-made catastrophes. Figures 3 and 4 show two methods that may be used to distribute DNA-rich materials to working and reserve nodes. In the first example (Fig. 3), materials might be collected in triplicate by a CENARGEN expedition and one of the replicates mailed to the Vavilov Institute (where it would be cryostored as a reserve node), another replicate might be mailed to the Gene Library (where it would be cryostored as the second reserve). CENARGEN would then take the third replicate home to the working node. This method would aid in assuring that the samples actually get sent to the reserve nodes, but it would be critical that considerable identification and label making be done in the field. A second method, shown in figure 4, would be to take all 3 replicates back to CENARGEN and then send the replicates to the Vavilov Institute and the Gene

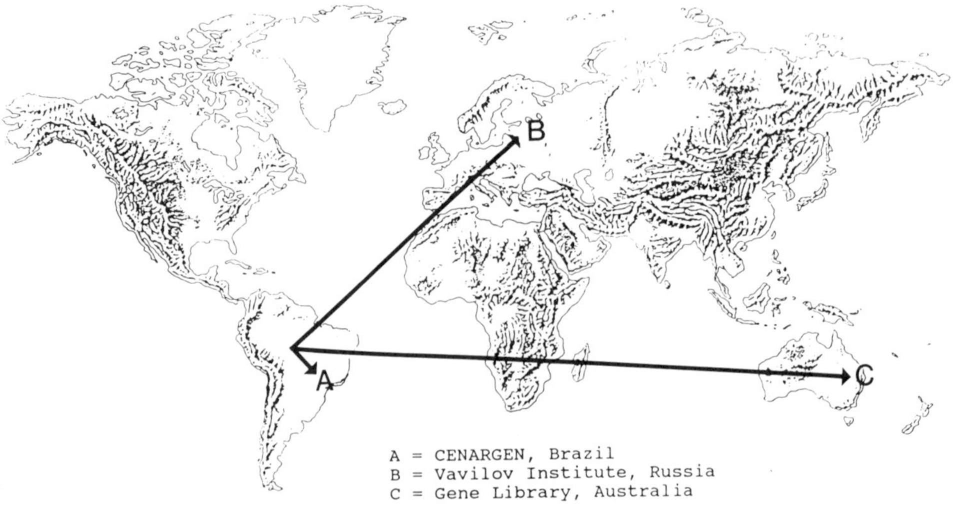

Fig. 3. Hypothetical example of a triplicate collection by CENARGEN in Brazil. The DNA-rich materials are placed in silica gel or drierite for interim preservation and then two replicates sent to reserve nodes at (B) Vavilov Institute and (C) the Gene Library. The third replicate would be taken to the working node (for this accession) at CENARGEN (A).

Library. The plant materials (in silica gel) could be stored in a freezer until the identification and field notes have been accomplished and then shipped in quantity with other samples in off-season periods. No doubt other strategies will be developed with experience. Several general recommendations came from the task groups (Adams and Adams, 1991) and these include:

a. DNA should be extracted from cryo-preserved DNA-rich materials only when the DNA is needed. Delaying the extraction has the advantage of letting technology catch up, so advanced techniques can be used as they become available.

b. Working nodes should generally be an existing organization with adequate biochemical expertise and have an associated herbarium. Having an herbarium on site would not be required but a very close, local (in the city) association with a recognized herbarium (Holmgren et al., 1990) is required.

c. For the working as well as reserve nodes, it is necessary to have a strong institutional commitment, not just a personal commitment, in order that the collection be maintained in perpetuity not just for the lifetime of one person committed to the idea.

d. Consideration should be made concerning the availability of dependable electricity and liquid nitrogen in determining the feasibility of establishing a node.

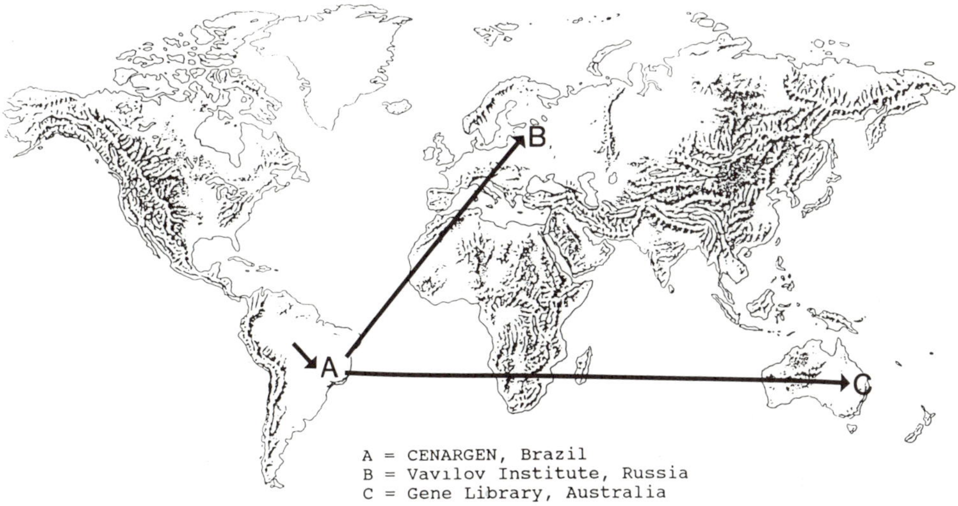

Fig. 4. Hypothetical example of a triplicate collection by CENARGEN in Brazil. The DNA-rich materials are placed in silica gel or drierite for interim preservation and taken to the working node (A, CENARGEN). From CENARGEN (A), two replicates would be sent to reserve nodes at (B) Vavilov Institute and (C) the Gene Library. This procedure would minimize the amount of paperwork needed in the field.

e. Considerable interest was shown in the concept of storing composite DNA samples (e.g., a composite of DNA from all the legumes in a region, to be used for screening or retrieval of unusual genes).

f. The need for computer and data base compatibility was expressed. Given the number of flat file and relational data bases that are compatible with dBASE, it would seem that dBASE compatibility would be desirable. No consensus was reached in regards to this nor on the use of a flat file vs. relational data base. It was felt that the critical issue at present was to begin collecting DNA-rich materials.

SCOPE OF PLANT COLLECTIONS

The task group given this assignment felt that there is a need for an initial focus rather than random collections and that economically useful plants should be given some priority (Adams and Adams, 1991). However, this priority would not include the major crop plants of commercial usage that are widely cultivated (e.g. maize, rice, wheat, etc.), but rather those indigenous species that are tended and/or otherwise used by local people.

One problem with giving a priority to species is that field collecting then becomes 'plant hunting' trips, which tend to be very expensive. It would seem that the cheapest and most practical way to preserve the largest percentage of plant genes would be to utilize the current (and additional) floristic collectors (such as those of the Missouri Botanical Garden, Royal Botanic Gardens, etc.), who are already in the field and familiar with the vegetation in the region. The collections of DNA-rich material (leaves) could be done with little additional effort when specimens are collected.

DNA COLLECTING PROCEDURES

DNA collectors should be considered the same as all other plant collectors. Consequently they should (Adams and Adams, 1991):

a. Voucher all collections in recognized herbaria (i.e., listed in Index Herbariorum, ed. 8)

b. Provide proper label information as to the locality, habitat, etc. for each plant collected.

c. Follow all procedures concerning permits, convenios, and deposition of duplicate vouchers in the country of origin.

d. Collect leaf samples and pack them in desiccants (see Adams et al., 1991) immediately (the same day). Leaves are of value as simple long-term storage.

e. In the case of legumes, samples of root nodules should be taken if possible, but kept as a separate accession.

f. If a chemical treatment is used in the field, information should be provided concerning the method and some untreated leaves must be stored in desiccant (see d. above).

g. Fossil plants - When possible, fossil material should be included in DNA Bank-Net. In this case, when destruction of the source material occurs, documentation via photographs and fragments is necessary.

h. Some material may be accessioned from herbarium specimens under control of local curators using current methods of DNA extraction. Herbarium sheets should be marked if sampled for DNA. Herbarium specimens are limited in supply and their utility appears to be limited to material collected without chemical preservation. Material may be sampled directly from the sheet or the attached specimen envelope if it contains sufficient leaf material (ca. 0.1 - 0.5 g dry wt.) for DNA extraction.

INTERIM FIELD STORAGE OF DNA SPECIMENS

The problems associated with bringing back fresh or frozen materials can generally be overcome by specialists (ex. world-wide collections of fresh foliage of *Juniperus* for essential oil analyses and DNA by RPA). However, botanists doing floristic research will likely collect many of the specimens from tropical rare and endangered species. They often collect specimens from scores of different species in a single day. The bulk of the materials that they have to process and ship requires that any protocol for the collection of samples for specialized needs (ex. DNA storage / analyses), must be quick, simple and trouble-free. The generalist collector, working in tropical areas cannot be expected to preserve hundreds or thousands of collections for months under tropical conditions and then arrange transport through customs, all the while keeping the individual specimens frozen.

Fortunately, at least as far as DNA preservation is concerned, interim preservation in silica gel or drierite is an effective way to keep plant materials in the field and/or in transit for several months at ambient temperatures (Adams et al., 1991). Figure 5 shows genomic DNA from fresh spinach stored for up to 6 months still contains high molecular weight DNA (see Adams et al., 1991 for detailed discussion).

Protocol for field preservation of foliage

Drierite has a water capacity of 10 to 14 percent, but above %, the capacity varies inversely with temperature (W. A. Hammond Drierite Co.). One would not want to risk possible rehydration of leaves, so storage ratios should be based on the 6.6% capacity. In lab tests, silica gel absorbed 8.85% of its weight of water after exposure to 100% humidity for 16 h at 22°C. We have found that plant materials contain as much as 92% moisture, so a useful approximation would be to assume the plant is mostly water and use 16 to 20 times the fresh leaf weight for the drierite or silica gel component.

Now that inexpensive ($100 USD) battery powered, portable balances are available, one could take a supply of jars that hold (for example) 100 g of silica gel and then weigh out 5 g of fresh leaf material and add it to the jar along with silica gel (or drierite). We have found that air dried leaves (suitable for herbarium vouchers) generally contain from 10 to 15% water. Using a robust value of 20% water for air dried leaves, one can weigh out 5 g of air dried leaves (5 g x 20% = 1 g water) or 1 g fresh leaves per 20 g of silica gel. This procedure may seem time consuming, but in practice, we merely do a quick

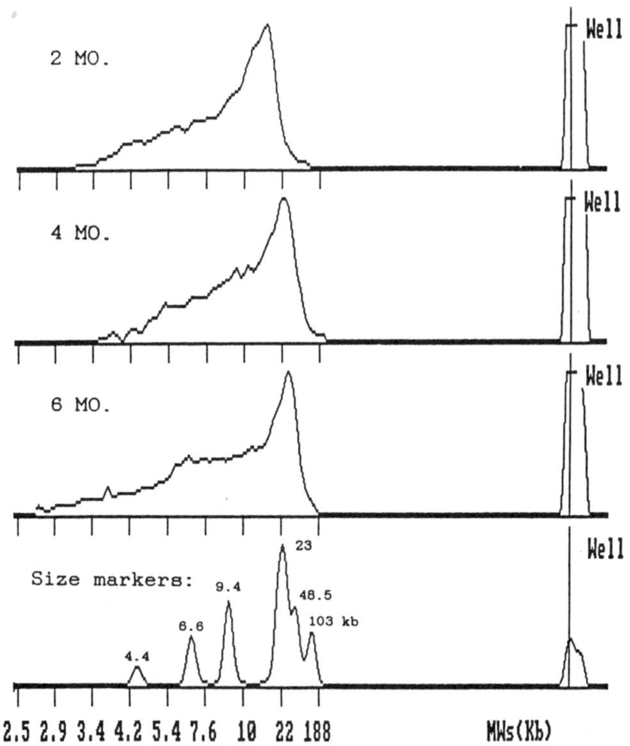

Fig. 5. Video densitometer scans of FIGE of genomic DNA from fresh spinach leaves stored in silica gel at 37°C for 2, 4 and 6 months (from Adams et al., 1991). Notice the gradual increase in lower molecular weight DNA (the peak tailing to the left). Size markers are: lambda DNA cut with HindIII (2.3, 2.0 and 0.5 kbp bands not shown); Lambda DNA (48.5 kbp) and T5 DNA (103 kbp).

check on the leaf area needed to give approximately 1 g (fresh leaves) or 5 g (dried leaves) and then just use that amount of leaf area. For example, for spinach, a 2 cm x 4 cm fresh leaf area weighs about 1 g. So, one can just cut the leaves into roughly 2 cm x 4 cm squares and add one square to 20 g of silica gel. For succulent leaves, a slightly different protocol may be used. Liston et al. (1990) removed succulent leaf material after 24 hours in drierite and placed it in fresh drierite. A note of caution is necessary concerning field drying of specimens for subsequent silica gel/drierite storage. We have experienced difficulty obtaining DNA from leaves dried at temperatures higher than about 55°C. In very rainy conditions where high drying temperatures (from butane stoves, for example) are used to dry specimens, it would seem advisable to merely blot leaves free of surface moisture and then place the fresh leaf material directly into silica gel or drierite. Liston et al., (1990) took 2-5 g of plant tissue and wrapped it in tissue paper to prevent it from fragmenting, then placed it in a 125 ml Nalgene bottle, 1/3 prefilled with drierite (with blue indicator crystals), and then filled the bottle (2/3) with additional drierite.

Plastic bottles are probably to be preferred to glass, to avoid breakage in transit. Using clear jars allows one to check the indicating crystals without opening the jar. The lids should be sealed with vinyl tape to insure against moisture leakage. The use of parafilm to seal containers is not recommended, as we have found it to come loose at 37°C (and of course, at tropical temperatures!).

Silica gel and drierite do differ in one characteristic that may be a consideration. We have found that silica gel can be dried (recharged) at 100°C for 24 h but drierite must be dried at much a higher temperature (200°C). In addition, we could easily dry silica gel in a microwave oven, but were unable to dry (recharge) drierite in a microwave oven. If the desiccant gets wet before use, silica gel appears to be much easier to dry. Silica gel is used in large quantities for flower drying and, thus, may be cheaper, depending on the source. Both drierite and silica gel could be recharged for reuse on subsequent trips, but one should be very careful to remove any leaf fragments. If the materials are to be checked through customs, it is useful to have a small container of silica gel/ drierite that you can open and show the customs agents. A demonstration that the blue indicator crystals will turn pink when you breath on or moisten them is helpful in convincing the customs officials to **not** open your sealed specimen jars.

FUTURE RESEARCH

The vast resources of dried herbarium specimens may hold considerable DNA that would be suitable for PCR. It seems likely that the integrity of DNA would decrease with the age of specimens. Because there are many types of herbaria storage, preservation and collections, there is a need for systematic investigations of the effect of modes of preparation, collection and storage on the integrity of DNA in the world's major holdings.

One of the major concerns in storing DNA from extinct species is the limited amount of DNA available for distribution. A general process by which the DNA could be immobilized and then specific genes or oligonucleotides amplified is needed. Figure 6 depicts the immobilization of genomic DNA onto nylon as described by Kadokami and Lewis (1990) for cDNA from spiders (*Nephila clavipes*). Amplification would then involve removing the membrane with the bound DNA from cryo-storage (Fig. 7) and amplifying the desired gene, washing away the primers and placing the bound DNA back into cryostorage. Although Kadokami and Lewis (1990) reported successful PCR amplification of membrane bound cDNA, we have not been able to extend their work to genomic plant DNA. Additional research is needed in this area.

Research is needed to amplify the entire genome DNA of a species. Some modification of the GAWTS (Genomic Amplification with Transcript Sequencing, Sommer et al., 1990) type protocol (Fig. 8) needs to be developed for eventual supplementation of DNA reserve stocks and obviate the need for replenishment from outside sources.

Concluding remarks

DNA Bank-Net is an association that fills a needs for a professional organization that would function initially as a lead organization and superstructure. The association brings together capabilities and expertise, produces a newsletter, and coordinates DNA banking activities.

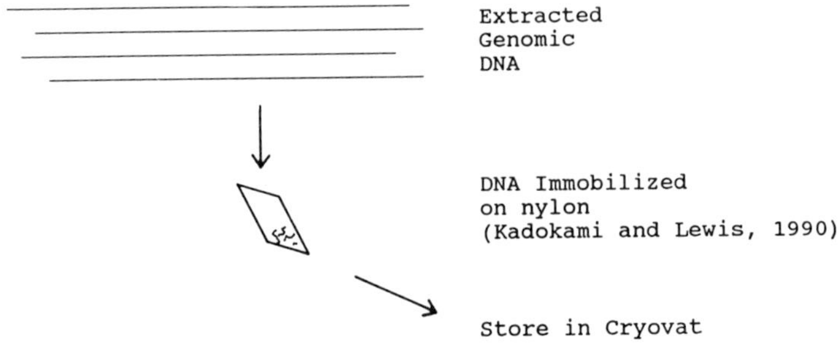

Fig. 6. Scheme for the immobilization of genomic DNA (after Kadokami and Lewis, 1990 and Adams and Adams, 1991).

Technical workshops need to be conducted in order to bring researchers together and develop specific techniques and protocols for DNA extraction, amplification and storage. DNA Bank-Net supports the concept of using license agreements and/or contracts that insure that there is a flow of money back to the countries and/or institutions when commercialization of protected germplasm is achieved. This will guarantee the uninhibited exchange of germplasm for scientific purposes, so that scientific achievements are not obstructed.

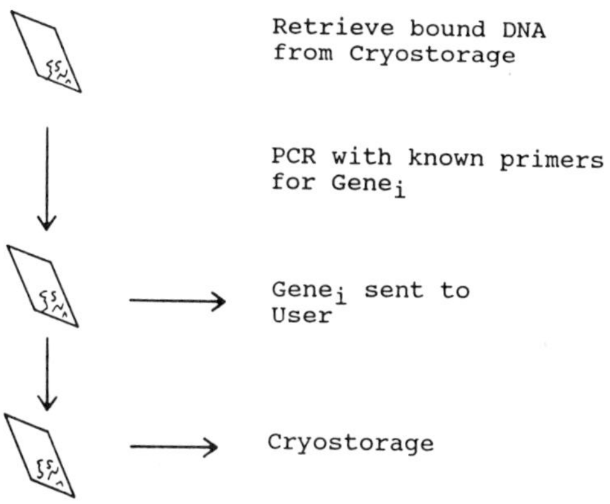

Fig. 7. Conceptual procedure for the amplification of immobilized genomic DNA (after Kadokami and Lewis, 1990 and Adams and Adams, 1991).

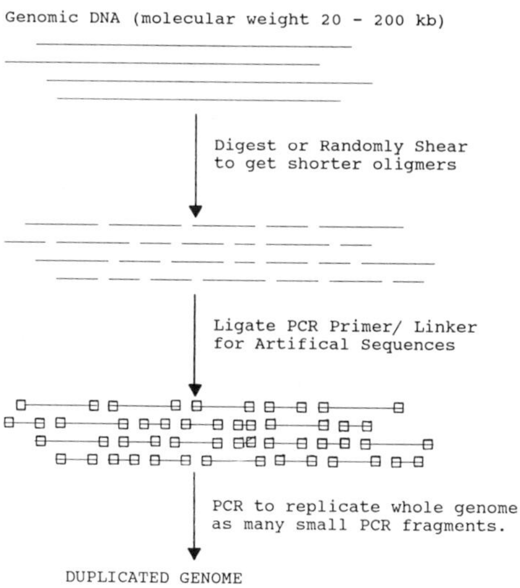

Fig. 8. Possible amplification of genomic DNA by use of a modified genomic amplification with transcript sequencing (GAWTS) method of Somer et al. (1990).

DNA Bank-Net should complement activities already being performed by different institutions, specifically, those working in the area of germplasm collection and conservation.

Technically, all the necessary expertise is available to begin the collection and storage of DNA from endangered species. Now, the vision is needed.

ACKNOWLEDGMENTS: DNA Bank-Net has been supported by funds from the Conservation, Food and Health Foundation, the Helen Jones Foundation and the Wallace Genetic Foundation.

REFERENCES

Adams, R.P., 1988, The preservation of genomic DNA: DNA Bank-Net, AIBS Meeting, Aug., Univ. of California, Davis, CA.

_____, 1990, The preservation of Chihuahuan plant genomes through in vitro biotechnology: DNA Bank-Net, a genetic insurance policy, Pp. 1-9. *in:* "Third Symposium on Resources of the Chihuahuan Desert Region," A.M. Powell, R.R. Hollander, J.C. Barlow, . McGillivray and D.J. Schmidly (eds.), Printech Press, Lubbock, TX.

_____, Adams, J.E., 1991, "Conservation of plant genes: DNA banking and in vitro biotechnology," Academic Press, NY.

_____, Do, N., Chu, G-L., 1991, Preservation of DNA in plant specimens by dessication, Pp. 133-150, *in:* "Conservation of plant genes: DNA banking and *in vitro* biotechnology," R.P. Adams and Adams (eds.), Academic Press, NY.

Barton, K.A., Whiteley, H.R., Yang, N., 1987, *Bacillis thuringiensis* d-endotoxin expressed in transgenic *Nicotiana tabacum* provides resistance to Lepidopteran insects, *Plant Physiol.* 85: 1103-1109.

Booth, W., 1987, Combing the earth for cures to cancer, AIDS, *Science* 237: 969-970.

Della-Cioppa, G., Bauer, S.C., Taylor, M.L., Rochester, D.E., Klein, B.K., Shah, D.M., Fraley, R.T., Kishore, G.M., 1987, Targeting a herbicide resistant enzyme from *Escherichia coli* to chloroplasts of higher plants, *BioTechnology* 5: 578-584.

Fischhoff, D.A., Bowdish, K.S., Perlak, F.J., Marrone, P.G., McCormick, S.M., Niedermeyer, J.G., Dean, D.A., Kusano-Kretzmer, K., Mayer, E.J., Rochester, D.E., Rogers, S.G., Fraley, R.T., 1987, Insect tolerant transgenic tomato plants, *BioTechnology* 5: 807-813.

Hawkes, J.G., 1979, Genetic poverty of the potato in Europe, Pp. 19-27. *in:* "Proceeding of the Conference: Broadening the Genetic Base of Crops," A.C. Zeven and A.M. van Harten (eds.), Wageningen, Netherlands, 3-7 July, 1978. PUDOC, Wageningen.

Holmgren, P.K., Holmgren, N.H., Barnett, L.C., 1990, "Index herbariorum. Part I. The herbaria of the world," ed. 8., *Regnum Veg.* 120.

Kadokami, Y., Lewis, R.V., 1990, Membrane bound PCR, *Nucleic Acid Res.* 18: 3082.

Lefebvre, D.D., Miki, B.L., Laiberte, J-F., 1987, Mammalian metallothionein functions in plants, *BioTechnology* 5: 1053-1056.

Liston, A., Rieseberg, L.H., Adams, R.P., Do, N., Zhu, G-L., 1990, A method for collecting dried plant specimens for DNA and isozymes analyses, and the results of a field test in Xinjiang, China, *Ann. Mo. Bot. Gard.* 77: 859-863.

Maiti, I.B., Hunt, A.G., Wagner, G.J., 1988, Seed-transmissable expression of mammalian metallothionein in transgenic tabacco, *Biochem. & Biophys. Res. Comm.* 150: 640-647

Murai, N., Sutton, D.W., Murray, M.G., Slightom, J.L., Merlo, ., Reichert, N.A., Sengupta-Gopalan, C., Stock, S.A., Barker, R.J., Kemp, J.D., Hall, T.C., 1983, Phaseolin gene from bean is expressed after transfer to sunflower via tumor- inducing plasmid vectors, *Science* 222: 476-482.

Nelson, R.S., McCormick, S.M., Delannay, X., Dube, P., Layton, Anderson, E.J., Kaniewska, M., Proksch, R.K., Horsch, R.B., Rogers, S.G., Farley, R.T., Beachy, R.N., 1988, Virus tolerance, plant growth, and field performance of transgenic tomato plants expressing coat protein from tobacco mosaic virus, *BioTechnology* 6: 403-409.

Newmark, P., 1987, Trypsin inhibitor confers pest resistance, *BioTechnology* 5: 426.

Plucknett, D.L., Smith, N.J.H., William, J.T., Murthi Anishetty, 1987, "Gene Banks and The World's Food," Princeton University Press, Princeton, New Jersey.

Raven, P.H., 1987, Forests, people, and global sustainability, Keynote Address, National Audubon Society Biennial Convention, Western Washington University, Bellingham, WA.

_____, 1988, Tropical floristics tomorrow, *Taxon* 37: 549-560.

Redden, R.J., Singh, S.R., Luckefahr, M.J., 1984, Breeding for cowpea resistance to Bruchids at IITA, *Protection Ecol.* 7: 291-303.

Sommer, S.S., Sarkar, G., Koeberl, D.D., Bottema, C.D.K., Buerstedde, J-M., Schowalter, D.B., Cassady, J.D., 1990, Direct sequencing with the aid of phage promoters, *in:* "PCR Protocols," M.A. Innis, D.H. Gelfand, J.J. Sninsky and T.J. White (eds.), Academic Press, San Diego.

Shulman, S., 1986, Seeds of controversy, *BioScience* 36: 647-651.

Ullstrup, A.J., 1972, The impacts of the southern corn leaf blight epidemics of 1970-1971, *Annual Review of Phytopathology* 10: 37-50.

U.S. Congress, 1987, "Technologies to maintain biological diversity," Office of Technology Assessment, Washington, D.C.

Appendix I. Individuals/institutions that have expressed an interest in DNA Bank-Net.

Dr. Robert P. Adams,Plant Biotechnology Center, BU Box 97372
Baylor University, Waco, TX 76798 USA

Dr. Daniel K. Abbiw, Botany Department, Box 55, University of Ghana,
Legon, Ghana, West Africa

Drs. Lucia Atehortua/Ricardo Callejas, HUA, Department of Biology,
University of Antiquia, Medellin, Colombia

Dr. Luiz Antonio Barreto de Castro, CENARGEN/EMBRAPA, Parque Rural,
CP 102372, Ave. W5 Norte, W 70770, Brasilia DF, Brazil

Dr. M.D. Bennett, Director, Jodrell Laboratory, Royal Botanic Garden, KEW,
Richmond, Surrey TW9 3AB, England

Dr. Anthony H. D. Brown, Division of Plant Industry, CSIRO
GPO Box 1600, Canberra, ACT 2601, Australia

Prof. Cheng Xiongqying, Institute of Nuclear Agricultural Sciences,
Zhejiang Agricultural University, Hangzhou, 310029 China

Prof. Chu Ge-lin, Institute of Botany, Northwest Normal College,
Lanzhou, Gansu, China 730070

Dr. Robert E. Cook, Arnold Arboretum, Harvard University,
125 Arborway, Jamaica Plain, Mass. 02130-3519

Dr. Marshall Crosby, Asst. Director, Missouri Botanical Garden,
2315 Tower Grove Ave., St. Louis, MO 63166

Dr. Thomas Elias, Director, Rancho Santa Ana Botanic Garden,
1500 North College Ave., Claremont, CA 91711

Prof. J. Eloff, Director, National Botanical Institute,
Private Bag X101, Pretoria, 0001 South Africa

Dr. Z. O. Gbile, Director, Forestry Research Institute of Nigeria,
Private Mail Bag 5054, Ibaden, Nigeria

Dr. David Giannasi, Dept. of Botany, University of Georgia,
Athens, GA 30602 USA

Dr. David Given, Department of Science and Industrial Research
Botany Division, Private Bag, Christchurch, New Zealand

Dr. Chaia C. Heyn, Dept. of Botany, The Hebrew University,
Jerusalem, 91904, Israel

Dr. Vernon Heywood, IUCN Conservation Monitoring Center,
53 The Green, Kew, Richmond, Surrey TW9 3AA, England

Dr. Toby Hodgkin, Research Officer, Genetic Diversity IBPGR,
c/o FAO of the UN, Via delle Sette Chiese 142, 00145 Rome, Italy

Prof. Hu Zhong, Professor of Plant Biochemistry, Kunming Institute of Botany,
The Academy of Sciences of China, Heilongtan, Kunming, Yunnan, China

Dr. Kunio Iwatsuki, Botanical Gardens, Faculty of Science,
University of Tokyo, 3-7-1 Hakusan, Bunkyo-ku, Tokyo 112, Japan

Prof. Mupinganayi Kadiakuida, Director General, CARI, B.P. 16513,
Kinshasa, Republic of Zaire

Dr. S. L. Kapoor, Cytogenetics Lab, National Botanical Research Institute,
Rana Pratap Marg., Lucknow 226 001, India

Prof. Lin Zhong-ping, Division of Plant Molecular Biology,
Institute of Botany, Academia Sinica, Beijing 100044, China

Dr. Ma Cheng, Chief Engineer, Bureau of Bio-Sciences and Biotechnology,
Chinese Academy of Sciences (Academia Sinica),
32 San Li He Road, Beijing, China

Dr. Lydia Makhubu, University of Swaziland, Kwaluseni Campus
P/Bag, Kwaluseni, Swaziland

Dr. John S. Mattick, The Gene Library, Centre for Molecular Biology and
Biotechnology, University of Queensland,
St. Lucia, Queensland, QLD 4072 Australia

Dr. James Miller, Missouri Botanical Garden,
2315 Tower Grove Ave., St. Louis, MO 63166

Dr. Titus K. Mukiama, Department of Botany, University of Nairobi,
Chiromo, P.O. Box 30197, Nairobi, Kenya

Dr. F. Ng, Deputy Director, Forestry Research Institute Malaysia (FRIM),
Karung Berkunci 201, JLN FRI Kepong, 52109 Kuala Lumpur, Malaysia

Dr. Mohinder Pal, Head, Cytogenetics Lab, National Botanical Research
Institute, Rana Pratap Marg., Lucknow 226 001, India

Dr. Bart Panis, Laboratory of Tropical Crop Husbandry, Catholic University of
Leuven, Kardinaal Mercierlaan 92. B-3001 Heverlee, Belgium

Dr. Ghillean T. Prance, Director, Royal Botanic Garden, KEW,
Richmond, Surrey TW9 3AB, England

Dr. Steve Price, Industrial Liaison Officer, Office of Intellectual Property,
Iowa State University, Ames, Iowa, 50010

Dr. Mukunda Ranjit, Assistant Pomologist, Fruit Development Division,
Kirtipur, Kathmandu, Nepal

Dr. W. Roca, Head, Bio-Tech Unit, CIAT, AA6713, Cali, Colombia

Dr. Phillip Stanwood, National Seed Storage Lab, USDA,
Colorado State University, Ft. Collins, CO 80523

Dr. Dennis Stevenson, The New York Botanical Garden,
Bronx, New York 10458-5126

Dr. Peter Strelchenko, N. I. Vavilov Institute of Plant Industry,
42, Herzen Street, 190000, Leningrad, USSR

Prof. Thavorn Vajrabhaya, Dept. of Botany, Faculty of Science,
Chulalongkorn University, Bangkok 10500, Thailand

Dr. Victor M. Villalobos, Director, Programa Mejoramiento Cultivos Tropicales,
CATIE, Turrialba 7170, Costa Rica

Prof. Luz Ma. Villarreal de Puga, Instituto de Botanica, University de
Guadalajara, Apartado 139, Las Agujas, Nextipac, Zapopan Jalisco, Mexico

Dr. Melaku Worede, Plant Genetic Resources Center,
P.O. Box 30726, Addis Ababa, Ethiopia

Prof. Zheng Sijun, Dept. of Agronomy, Zhejiang Agricultural
University, Hangzhou, 310029, Zhejiang, China

THE EXPLOITATION OF GENETIC RESOURCES

Peter R. Day

Center for Agricultural Molecular Biology
Rutgers, The State University of New Jersey
New Brunswick, NJ 08903

INTRODUCTION

Modern agriculture depends on a flow of improved forms of crop plants. These are fashioned by breeders from germplasm which provides the genetic information needed to increase yield and quality and add resistance to pests, diseases and adverse environmental factors such as drought, temperature extremes and toxic soils. In the industrialized countries genetic resources are maintained and evaluated in collections supported by public funds. The Consultative Group for International Agricultural Research (CGIAR) supports international, regional and some national collections for the benefit of industrialized and developing countries. Among these are the collections of the International Agricultural Research Centers (IARCS) one of which, the International Board for Plant Genetic Resources, plays a central coordinating role. While agriculture has always been concerned with maximizing production at the lowest possible monetary cost, other factors, such as the environment and consumer health, have begun to play a part in determining agricultural strategies in recent years. In the developing world agricultural production is geared first to satisfying the demands of human hunger, and then to exports that generate foreign exchange to improve the standard of living of the population. However, here also pressure to maintain environmental quality is growing. There are two views of the exploitation of genetic resources. One is that it represents the utilization of a natural resource. The other sees it as an unjust or improper use of another's resources for one's own profit and advantage. During the last 10 years this latter view has become more and more apparent in the exchanges that have come to be known as the North-South debate. This view of exploitation now poses a serious threat to the free exchange of germplasm and therefore to its general utilization. In this paper I review some of the factors responsible for it and

Gene Conservation and Exploitation
Edited by J.P. Gustafson *et al.*, Plenum Press, New York, 1993

the problems and opportunities raised by recent developments in crop plant germplasm exploitation through biotechnology.

Two sources were particularly useful; the series of three Keystone Dialogues on Plant Genetic Resources provided a forum for extensive discussions between representatives of industry, government, public and private plant breeders and non-governmental organizations (NGOs). The second was a workshop to discuss what policies the IARCs should follow regarding the protection of intellectual property (Barton & Siebeck, 1991) and biosafety, held in September 1991 at the International Service for National Agricultural Research, in The Hague, Netherlands.

PROTECTING THE PROFIT FROM GENETIC RESOURCES

In the industrialized world new varieties of crop plants are produced by the public and private sector to satisfy the needs of farmers. In some 17 countries a system of plant breeders rights (PBR) protects the interests of the breeder. The resulting intellectual property protection (IPP) encouraged the development of a private plant breeding industry. Varieties that are shown, typically by a national testing authority, to be new, distinctive, uniform and stable, and to have value in cultivation and use, are registered. When they are sold by seed merchants a portion of the receipts is returned to the breeders as royalty payments. Although farmers have not been restricted from saving seed for their own use in subsequent seasons, there are restrictions on their selling this seed to other farmers, for planting without payment of royalties. Breeders are free to use varieties owned by other breeders as parents in their own breeding programs without permission, but require permission for repeated use, for example as a parent of an F_1 hybrid. In regions such as western Europe, where farmers commonly purchase up to 80% of cereal seed for planting each year, the royalty income sustains a vigorous plant breeding industry. However, in the U.S. and Canada, where up to 80% of wheat seed is home saved for replanting the following season, the relative financial returns are lower. In those crops that can be grown as F_1 hybrids protection is achieved by not revealing or releasing the inbred parents of the hybrids. In this way farmers must buy fresh F_1 hybrid seed each season. Public sector breeders, at breeding stations and university departments, have traditionally used public funds for the public good and have not sought financial profit. However, the introduction of PBR has enabled public sector breeders to earn royalty income for their organizations which may be used to support their breeding programs. Before the Cambridge Plant Breeding Institute was privatized in 1987, for each of the previous 4 years its varieties of agronomic crop plants earned royalties that exceeded its annual grant in aid to cover all operational costs for breeding and basic research. These royalties were returned to the U.K. treasury and were not directly applied to support the plant breeding programs that generated them. In 1991 the International Union for the Protection of New Varieties Convention (UPOV), the organization concerned with establishing and implementing PBR, negotiated a new convention that is not yet in force. One important change is the extension of the breeder's rights to the harvested material produced from non-authorized use of propagating material. Individual nations are now free to extend this right to farmers' seed propagated for their own use. Another is that an essentially derived variety, such as a somaclonal variant or a transformant carrying an added gene, is still subject to the rights of the original breeder. In other words, before the new variety can be protected by PBR or marketed, the owners of the original and the derived variety must agree on royalty sharing. An important feature of PBR has been the understanding that breeders may freely use protected materials belonging to other breeders as parents in crosses to generate new varieties. The advent of DNA transformation has led to the view that this freedom is no longer acceptable. As a consequence most organizations that are developing new forms through DNA transformation are protecting their interests by patenting the products, or the genes involved, or both. Patents commonly give rights for 20 years that enable the patent holder to stop others from exploiting the invention commercially. To receive patent protection the invention must satisfy several conditions. It must be new and not an obvious application of what is already known. It must be disclosed so that others can reproduce it, and it must relate to a technology in which patents are permitted. Also the scope of the protection granted should be in proportion to the invention (Keystone, 1991). Although some countries in western Europe do not recognize patents on living organisms, most countries with patent laws now allow the patenting of isolated genes that can be introduced into a new context to create a new product. This means that genes synthesized in the laboratory and genes from nature, both of which can be described by their DNA base sequences, are protectable. The latter are not protectable when they occur in their natural context. Plant varieties protected by patents cannot be freely used as parents in crosses. Patent owners allow their use in research but may control the commercial use of any products which incorporate the patented gene. The owner of the patented gene can thus control the marketing of the product grown from the patent protected variety in countries that recognize his rights. If it is a grain crop such as corn, wheat or rice, he might be able to control where the harvested grain is exported or marketed for food use.

EFFECTS OF INTELLECTUAL PROPERTY RIGHTS ON EXPLOITATION

Both PBR and patents are regarded as a means of protecting the substantial investment in the research and development needed to make discoveries and bring them to the market place. Although widely accepted in the industrialized world, some parts of the developing world see PBR and patents as a means whereby the industrialized nations control international trade and exploit the natural resources of the developing world. Some developing countries also see the patents held by multinational companies as a means of preventing them from adopting and benefiting from biotechnology. These criticisms have been aired at the FAO Commission on Plant Genetic Resources and in many publications from NGOs who have campaigned for broader and monetary recognition of the value of the indigenous germplasm in the developing world. An example is the principle of Farmer's Rights which calls for recognition of the past, present and future contributions made by farmers and indigenous peoples in the development of the cultivars and land races which are the ancestors of our current crop cultivars. These materials are a valuable source of genes for quality, agronomic performance, environmental adaptation and pest and disease resistance, and are an important part of each of the major germplasm collections. Making Farmers Rights analogous to Plant Breeders Rights, by returning some form of royalty to the original farmers or their heirs, is widely acknowledged to be impossibly difficult. However, schemes such as imposing a modest tax on royalties to create a global fund specifically for germplasm conservation have received a good deal of support. Not surprisingly they are unpopular with the seed industry. They also introduce the problem of devising a satisfactory method for distributing the funds to where they can do the most good in advancing germplasm conservation.

BIOTECHNOLOGY AND THE INTERNATIONAL CENTERS

The IARCs with mandated crops have established germplasm banks and have traditionally distributed both germplasm, and the products of their research programs, with no strings or conditions attached. Most of them have favored a so-called open door policy. If IARCs wish to use genetic materials from elsewhere that are patented, they are likely to have no choice other than to pass on to their clients in the developing world the onus of negotiating, with the owner of the patent, a license to use the improved material. IARCs with research programs that are generating patentable materials must also now consider whether or not to obtain patent protection to prevent others from doing so and restricting the availability of IARC inventions. Not unexpectedly there is some expectation on the part of donors to the CGIAR system that Centers should take the opportunity to generate income and reduce their dependence on donor support. A simplistic view of the problems that arise from patenting germplasm is that a developing country that does not recognize patent protection is free to use patented material from elsewhere. Since the patent holder by definition has no rights there, the patent is not infringed by such use. But should seeds, planting stock, or harvested products be exported to countries where the patent is recognized there will be major problems. For example, some farmers in Colombia grow without a license, varieties of cut flowers developed in other countries where they are protected. When these flowers are marketed in the U.S. and western Europe, where protection is recognized, they infringe the original owner's rights and will be subject to seizure. Many developing countries are beginning to explore how they can benefit from biotechnology. In the industrialized world negotiations to establish research and development collaborations depend on the partners establishing crosslicensing arrangements that give each of them competitive advantages in the market place. Countries, such as India, that have rejected adopting patent law, have no bargaining chips in negotiations with the industrialized countries. The absence of patent laws introduces other problems. With no patent protection for pharmaceuticals in India drug prices to the public are low compared with other countries. But as a consequence, there is little significant investment by the pharmaceutical industry, in research on nationally important medical problems peculiar to India. Without intellectual property protection this is judged to be unprofitable (John Barton, personal communication). There is a steadily mounting pressure on countries without IPP laws to enact them. This is likely to have both negative and positive effects on germplasm conservation. Many of the centers of origin of agronomic crops are in the developing world. The nations where they occur are likely to restrict access to their germplasm since they see it as a valuable natural resource. This will not only limit the flow of germplasm from the South to the North but will also impede the exchange of germplasm among the developing countries themselves. On the other hand there could be an increased investment by developing countries in conserving their germplasm. At the third Keystone Dialogue (1991) the view was strongly expressed that conservation should be supported by an international fund. Participation in such a fund should require that a nation makes its conserved germplasm available to others. Another question concerns the policies that germplasm banks should adopt in distributing materials. Some are considering statements, to accompany germplasm samples sent in response to requests, that reserve the right of the bank to claim a royalty return if the material is subsequently used commercially. This return could be shared with the nation where the germplasm originated, the bank using its share to offset some of its operational costs. Although the likelihood of substantial returns to germplasm banks is low the arrangement could ensure that both the country of origin and the bank could benefit from any spectacularly successful exploitations.

EFFECTS OF BIOTECHNOLOGY ON EXPLOITATION

The international germplasm banks are an important source of materials for all methods of genetic manipulation. However, the methods of DNA transformation are not as rapid as some had hoped nor are they user friendly at the present time. Germplasm evaluation and regeneration will continue to depend, for the foreseeable future, on conventional methods that are both labor intensive and expensive. This work is essential if we are to know what is in the banks and that it is alive. At the same time a major investment of research effort is required to produce useful transgenic forms. Although gene isolation and transformation will become cheaper and more efficient it is unlikely that germplasm banks will be used by biotechnologists to the extent they are used by some plant breeders. Screening procedures will be very important. For example the cowpea trypsin inhibitor gene, an important source of insect resistance, was discovered through research to find out why seeds of a certain cowpea accession were resistant to insect damage in storage (Lombardi-Boccia et al. 1991). The germplasm banks will continue to be useful sources of minor gene adaptive characters for manipulation by conventional plant breeding methods. The production of detailed maps of crop plant genomes, using restriction fragment polymorphisms (RFLPs) and randomly assigned probes made with the polymerase chain reaction (RAPD-PCR) as markers, will help in the identification of quantitative trait loci in breeding programs but is unlikely to be useful in the short to medium term for screening germplasm until the methods are more user friendly and inexpensive than is presently the case (Lande and Thompson, 1990). It has been said, with some justification, that many of our seed banks are seed morgues (Goodman & Hernandez, 1991). There is no question that living assemblies of genetic determinants, represented by viable seeds and plant material, are of far more practical use to breeders than DNA sequences stored in ethanol in a freezer or as symbols in a data bank. However, this should not blind us to the potential of recovering genetic information from otherwise dead materials such as dead seeds and herbarium specimens, and even fossils, by using PCR with DNA probes designed to recover specific sequences.

BIOSAFETY AND GERMPLASM EXPLOITATION

Does Exploitation Increase Genetic Uniformity?

One of the penalties of successful large-scale modern agriculture has been the introduction of monocultures over large areas of land. Monocultures increase the threat of devastating losses caused by pests and diseases (NAS, 1972). The genetic vulnerability consequent upon genetic uniformity leads in turn to a tread-mill of pesticide application and intensification of efforts to breed for resistance. As each measure is deployed on a large scale the selection pressure for the pest and pathogen populations to respond in kind grows. The outmoded genes and chemicals are replaced by new ones at greater and greater cost. How will biotechnology affect this? Breeding for resistance has always been an important part of pest and disease control and has helped greatly to reduce the level of pesticide usage for many crops. Unfortunately corn and cotton have not been among them - at least for the control of lepidopteran pests. Biotechnology should eventually make it easier to manipulate genes for resistance than by conventional breeding. The introduction of the Bt endotoxin gene into both corn and cotton promises control over the short term but is unlikely to survive long exposure as the sole source of genetic resistance. Many groups, including our own at Rütgers, are deploying Bt genes to control lepidopteran, coleopteran and dipteran pests of crop plants. The toxin product of the cry-2 gene, effective against lepidopteran larvae, was successful and long-lived as the biological pesticide BT because it is short-lived in nature. After application as a spray BT is washed off by rain and/or inactivated by sunlight. It does not usually persist long enough to select resistant target insects. The use of BT and related safe biological pesticides could be compromised by the appearance of resistance in the pest populations exposed to transgenic plants expressing these toxins. The history of new introductions suggest that there is indeed a risk that this will happen unless the gene's exposure in transgenic crops is controlled or modulated. This can be done by reducing the level of expression so that only partial control is obtained (Gould, 1989), limiting expression to the times or plant parts that are most in need of protection, or by using it in mixed populations so that it is not exposed alone (Gould & Anderson, 1991). The latter strategy has been advocated for controlling fungal pathogens with host genes for disease resistance (Wolfe, 1988). All of these methods reduce the intensity of selection for toxin resistant insects.

Some commentators are worried that the current interest in engineered herbicide resistance will encourage excessive use of these materials (Goldburg et al., 1990; Hobbelink,1991). There is of course the threat of harming the environment though excessive herbicide. There is also the prospect of selecting new resistant weeds, and resistance in weeds that were formerly susceptible, which render environmentally benign herbicides useless. However, judicious use of herbicide resistance offers important new opportunities for environmentally sound weed control especially in minor crops (Duke et al. 1991). Another misuse of herbicide resistance in rice is discussed below. It is clear that genetic engineering introduces the prospect of increased genetic uniformity. The genes for Bt endotoxins, cowpea trypsin inhibitor, some virus coat proteins, and glyphosate resistance, to name only some of the best known ones, are now being introduced, together with selectable markers such as kan^r, conferring resistance to the antibiotic kanamycin, and controlling elements like the cauliflower mosaic virus 35s promoter, into a number of different species. There has been much interest in introducing nitrogen fixation in non-legumes. Most plant scientists remain skeptical because of the difficulty of protecting nitrogenase from oxygen, and the energy cost of fixation which would reduce harvestable yield. Even so, when this breakthrough is made and constructs are available for introduction into a variety of crop plants to reduce their dependence on nitrogen fertilizer, their exploitation would lead to rapidly increasing uniformity for nitrogen fixation. Genetic uniformity is not necessarily bad in itself. However, like the products of conventional plant breeding programs, some genetically engineered constructs may have undetected flaws that are only expressed under unusual or special circumstances. These may include particular environmental pressures, or new or uncommon forms of pests and pathogens. Could such flaws result in agricultural catastrophes like the Southern corn leaf blight of 1969-70? This was due to the unusual sensitivity of corn hybrids, made up with Texas cytoplasmic male sterile parents, to a new race of the fungal pathogen *Bipolaris* (*Helminthosporium*) *maydis*. Weather conditions also favored the rapid spread of the blight. An added risk with transgenic plants is that vulnerability may not be confined to one crop species but may be a feature of others that incorporate the particular genetic construct. If its potential flaws are known beforehand the risks and benefits of employing it in a wide range of crops can be weighed. The danger of genetic uniformity is that it enhances vulnerability to undetected flaws. As with all forms of genetic vulnerability the remedy lies in maintaining diversity.

GENETICALLY ENGINEERED ORGANISMS IN THE ENVIRONMENT

There is now general agreement that observance of sensible laboratory guidelines for containment have assured the public of the safety of recombinant DNA procedures. Over the last 15 years there have been no examples of hazards that have arisen that are peculiar to recombinant DNA technology. It is widely accepted that biosafety should be concerned with the product and not the process. However, there are others who do not share this view. The exploitation of genetic resources through biotechnology inevitably involves release to the environment. These releases may be in the form of farm crops but are expected also to include microorganisms designed to aid crop growth or protect them from parasites. The US, and several other countries, have adopted guidelines that call for careful review and safeguards designed to prevent both undesirable spread of the organism itself and horizontal transfer of genetic information to other plants. The Animal and Plant Health Inspection Service (APHIS) of the USDA has approved in the last 3 years, some 250 field trials of genetically engineered organisms all of which proceeded without incident. This accumulated experience is building expertise in assessing and evaluating risk. As both the regulators and the general public gain confidence this should eventually simplify the issue of permits and lead to appropriate and workable safeguards for the commercial introduction of transgenic crops and microorganisms. For field release the guiding principle is evaluation of the risk associated with the product. For example red rice is a major weed in some rice growing areas of the world. It can be controlled by using glyphosate prior to planting, or before emergence of the cultivated rice. The two species are interfertile and it is therefore very likely that the introduction of transgenic rice with resistance to glyphosate in these areas would in turn transfer the gene to red rice. As a result farmers would no longer be able to control red rice with glyphosate. A similar problem is posed by inserting the Bt gene into corn to control earworm. In Mexico corn and teosinte freely intercross in farmers' fields. It is likely therefore that the Bt gene would move into the native teosinte population. However, there is some question of how serious, if at all, the consequences of this would be. Those who are concerned that the gene pool would be contaminated must acknowledge the extensive introgression with cultivated varieties that occurs now. Of greater concern is the fact that if there is strong selection for the gene in teosinte this could accelerate the appearance of insect biotypes resistant to Bt endotoxin that would render the gene useless in corn in those areas. Unfortunately Bt resistance is already known in the diamondback moth in Hawaii (Tabashnik et al. 1991) from large scale field applications of BT insecticide. These two examples emphasize the importance of careful use to assure long term benefits from genetic exploitation. Transgenic organisms are not unique in this respect. There are many examples of the profligate use of chemical pesticides, and resistance genes in conventional breeding programs. Used alone, as the sole method of control, and on a very large scale, they impose strong selection pressure on the populations of target organisms. They frequently succumb to the new biotypes they select.

GLOBAL GUIDELINES FOR RELEASE

In the US environmental release is relatively uncomplicated by the prospect of hybridization with the indigenous flora. Most of the major agronomic crops were introduced during the development of modern agriculture and have no wild relatives. In the developing world this is not the case. We are already familiar with the displacement of land races and wild species by modern cultivars. The introduction of well adapted genetically engineered forms of crop plants to their centers of origin will likely further complicate current efforts to conserve natural variation in wild populations that may be useful to future generations. For example the introduction of engineered insect resistance in *Solanum tuberosum* is of great interest to CIP the IARC responsible for work on the potato, located in Lima, Peru. What impact would its release in the Andean potato region have on the native populations of Solanum spp. that are interfertile? Is the potential loss of resistance by uncontrolled adaptive changes in the insect population the only potential serious consequence or are there other changes that are more difficult to predict that should be considered before release is embarked upon? In these areas the further contamination of gene pools may well become a source of concern. Maize breeders would point to the fact that teosinte has not been used as a source of genetic variation in breeding. Most breeders are therefore skeptical about how important it is as a germplasm resource. However, it may be premature to make this judgment. It is impossible to predict the value of a gene pool in the future since we cannot know in advance what genes we may be looking for 20 - 200 years from now. The USDA - APHIS system of approvals and environmental assessments of each application depends on skilled and experienced staff. A system of this kind will be difficult for many developing countries to adopt since they often lack the bureaucratic infrastructure and trained personnel. Clearly the IARCs must take a lead in implementing biosafety procedures. At the Hague workshop it was agreed that at a minimum these should conform to their host country's regulations but that they should preferably follow the best practice in the industrialized world.

PRODUCT SAFETY

One of the most important potential applications of biotechnology is for the development of improved food crops. In the US the Food and Drug Administration (FDA) has the responsibility of assuring the safety of the food supply. The products of most conventional breeding programs are assumed to be risk free if they fall within the classification Generally Recognized as Safe

(GRAS) (Reitz & Caldwell 1976). Although a food manufacturer can decide for itself that a genetically engineered product is GRAS and use it without seeking FDA approval it runs the risk that FDA may not agree and challenge the product on the grounds that it contains an unapproved food additive. For example this might be the product of an introduced gene such as Bt endotoxin. A recent report (IFBC, 1990) examined the problems that face regulatory agencies such as the FDA and made several proposals for dealing with them. First the new product must be shown from laboratory tests or animal feeding trials to be harmless even when consumed at levels that are considerably in excess of normal intake. The report also suggests that new products should be evaluated to show that levels of toxic compounds known to occur in conventionally bred varieties are not increased as a result of unintended genomic changes introduced by transformation. Transgenic forms of tomato and potato would be tested for unusual levels of alkaloids in the fruits or tubers. One genetic engineering company has already petitioned the FDA for permission to incorporate the selectable marker gene that codes for the enzyme neomycin phosphotransferase II. This gene confers resistance to the antibiotic kanamycin. Coupled to the genes that are of agronomic interest, it enables the biotechnologist to select transformed plant cells in culture. Important policy issues remain to be resolved. If it is generally accepted that safety should be concerned with the product and not the process it would be very unfortunate if the costs of product regulation were so high that transgenic food plants become uneconomical. Food processors are also concerned as to whether or not the products will have to be labeled for the consumer. Biotechnology will enhance the current interest in "designer foods" with lower levels of carcinogens, and higher levels of compounds shown to protect experimental animals against carcinogens, all of which provide new challenges for germplasm exploitation. There is also great interest in using the techniques of molecular biology to enhance conventional methods of exploitation. The products of this kind of work should pose no new problems in product evaluation.

TWO RISKS

There are two sets of risks to agriculture and the environment associated- with biotechnology (Day, 1990). The first we have discussed in the context of biosafety. The second set of risks are those of not implementing this technology because we are afraid of its potential consequences. I believe this latter set of risks is the greater since a failure to pursue biotechnology could well compromise the future for agricultural productivity and enhanced environmental protection. As society continues to evolve we must also enlarge and improve the options available to preserve the quality of our lives and our environment.

ACKNOWLEDGEMENT

I am grateful to John Barton for his helpful comments and suggestions.

REFERENCES

International Agricultural Research Centers: What are the options? CGIAR Secretariat Paper pp. vi + 35.

Day, P.R. 1990. Two kinds of risks. pp. 6-11 in Risk Assessment in Agricultural Biotechnology: Proceedings of the International Conference. J.J.Marois G.E. Bruening eds. Univ. Cal. Press. Publication No. 1928.

Duke, S.O., Christy, A.L., Hess, F.D. & Holt J.S. 1991. Herbicide resistant crops. 24 pp. Council for Agricultural Science and Technology.

Goldburg, R., Rissler, J., Shand,H. & Hassebrook,C. 1990. Biotechnology's Bitter Harvest. Biotechnology Working Group, 73pp.

Goodman, M.M., & Hernandez, J.M. 1991. Latin America maize collections: a case for urgent action. Diversity, 7:8788.

Gould, F. 1989. Integrating biotechnology into agricultural systems: the need for more experimental agriculture at the farm level. pp 117-124 in Reform and Innovation of Science and Education: Planning for the 1990 Farm Bill, Committee on Agriculture, Nutrition, and Forestry United States Senate. US Govt Printing Office, Washington.

Gould, F. & Anderson, A. 1991. Effects of *Bacillus thuringiensis* and HD-73 delta-endotoxin on growth, behavior, and fitness of susceptible and toxin-adapted strains of *Heliothis virescens*. Environmental Entomology 20:30-38.

Hobbelink, H. 1991. Biotechnology and the Future of World Agriculture. 159pp Zed Books, Ltd., London and New Jersey.

International Food Biotechnology Council, 1990. Biotechnologies and food: assuring the safety of foods produced by genetic modification. Regulatory Toxicology and Pharmacology 12: Number 3 part 2; xxii + 196 pp.

Keystone Center, 1991. Global initiative for the security and sustainable use of plant genetic resources. Keystone International Dialogue Series on Plant Genetic Resources, Oslo plenary session. Pp vii + 43.

Lande, R. & Thompson, R. 1990. Efficiency of marker-assisted selection in the improvement of quantitative traits. Genetics 124:743-756.

Lombardi-Boccia, G., Carbonaro, M. & Carnovale, E. 1991. Trypsin and chymotrypsin inhibitors from a wild species and a domestic species of cowpea (*Vigna unguiculata*). Lebensmittel-Wissenschaft Technologie, 24: 370-375

NAS, 1972. Genetic Vulnerability of Major Crops. National Academy of Sciences, Washington,D.C. 307 pp.

Reitz, L.P. & Caldwell, B.E. 1974. Breeding for safety in field crops. Pp 1-5 in The effect of FDA regulations (GRAS) on plant breeding and processing, C.H. Hansen ed. Special Publication, Crop Sci. Soc. Amer.

Tabashnik, B.E., Finson, N. & Johnson, M.W. 1991. Managing resistance to *Bacillus thuringiensis*: lessons from the diamondback moth (Lepidoptera: Plutellidae). Journal of Economic Entomology 84: 49-55.

Wolfe, M.S. 1988 The use of variety mixtures to control diseases and stabilize yield. Pp 91-100 in Breeding Strategies for Resistance to the Rusts of Wheat. N.W. Simmonds & S. Rajaram eds. Mexico,D.F., Centro Internacional de Mejoramiento de Maiz y Trigo.

MOLECULAR EVALUATION OF PLANT GENETIC RESOURCES

Michael T. Clegg

Department of Botany and Plant Sciences
University of California
Riverside, CA 92521

INTRODUCTION

A central problem of biology is the description, classification and management of organic diversity. In one form or another this theme pervades biological science from the ecosystem level down to the population level. Diversity also has a central role in the science of genetics which originated from attempts to uncover the rules that govern the transmission of phenotypic variations between generations. Every student of introductory biology learns that the Mendelian method requires alternate allelic forms of a gene for any genetic analysis. Similarly, plant and animal improvement programs are absolutely dependent on useful genetic variants to achieve their goals of greater agricultural productivity (Brown et al., 1988; Day et al., 1991).

Biological diversity on Earth is being lost at an ever accelerating pace, owing to the pressures of an expanding human population (Ehrlich and Wilson, 1991; Soule, 1991). This loss has manifold consequences that range from global environmental change down to the economic costs associated with the loss of valuable plant and animal genetic resources. Serious efforts were mounted in the 1960s to conserve the genetic resources of crop plants and their wild relatives through the establishment of a global system of gene banks (Frankel and Bennett, 1970). These efforts have lead to the collection and maintenance of hundreds of thousands of accessions that represent broad geographic samples of most of the major cultivated plants and their wild relatives (Cohen et al., 1991). The wealth of collections that have been amassed in the past three decades have proved to be a valuable resource for plant breeders. Despite this success, genetic assessment of gene bank collections and the characterization of genetic diversity in standing populations of valuable plant materials has proceeded at a much slower pace.

Gene Conservation and Exploitation
Edited by J.P. Gustafson *et al.*, Plenum Press, New York, 1993

A wide variety of methods are available for genetic assessment. These range from various phenotypic evaluations through to sophisticated molecular techniques. The last twenty years has seen the development of a remarkable number of molecular techniques that permit the resolution of genetic differences at the level of individual nucleotide differences in DNA molecules. The goal in this chapter is to discuss the limitations and the strengths of these new molecular approaches to genetic evaluation. To provide a context for the evaluation of these methods I will first consider empirical and theoretical results from population genetics where the study of genetic diversity has a long history.

WHY UNDERTAKE GENETIC ASSESSMENT?

The economic justification for the conservation of plant genetic resources is based on the future utility of genetic variants for crop plant improvement. The precise nature of these future needs is often unknown. Given this uncertainty about future utility, conservation strategies must be based on general principles and not on the requirements of specific cases. An immediate corollary of this argument is that conservation strategies should aim to collect and preserve as wide a sample of the extant genetic diversity in the target species as is feasible. Clearly genetic assessment is of great value in the efficient design of collections programs (Brown and Clegg, 1983). Once collections are deposited in a gene bank, a host of management issues also require genetic information. These include guarding against the inadvertent contamination of accessions, evaluating the level of redundancy in collections and establishing core collections.

THEORETICAL BACKGROUND

The source of all genetic variations is mutation. We know from classical population genetic theory that all mutations are destined to be either lost or fixed in finite populations. The probability of loss or fixation and the expected time until either event occurs is a function of the effective population size and the strength and pattern of selection. The "neutral case" where selection is absent ($s=0$ where s is the selective coefficient and measures selection intensity) is the simplest and best studied. For a diploid population of effective population size N and μ mutation rate per locus per generation, the probability of ultimate fixation of a new mutation is $1/2N$. The expected time to fixation (t), given that fixation will ultimately occur, is approximately $4N$ generations (Crow and Kimura, 1970). For a selected gene, where the fitness of the heterozygote is intermediate between the two homozygotes (genic selection), the ultimate probability of fixation is approximately $2s$ (for s small) and the expected time to fixation, given that fixation will ultimately occur, is

(2/s)ln(2N) generations. It is instructive to calculate t for various choices of N and s. Table 1 presents these calculations for reasonable choices of N and s. Evidently the selected gene sweeps through the population very rapidly, while the neutral gene drifts in gene frequency for a very long period of time before fixation occurs.

While selectively favored genes are expected to reach fixation quickly, the probability of ultimate fixation may be small. For example, a gene with a 1% selective advantage will only be fixed 2% of the time under the genic selection model. Most favored genes are destined to be lost very

Table 1. The probability of fixation of a new mutation and the conditional time to fixation, given that fixation will ultimately occur, as a function of the effective population size (N) and the intensity of selection (s).

Selection	N	Fixation probability	Expected time to Fixation
s=0	10,000	0.0005	40,000
s=0.01	10,000	0.02	2,119
s=0.05	10,000	0.10	396
s=0	100,000	0.00005	400,000
s=0.01	100,000	0.02	2,441
s=0.05	100,000	0.10	488
s=0	1,000,000	0.0000005	4,000,000
s=0.01	1,000,000	0.02	2,902
s=0.05	1,000,000	0.10	580

quickly owing to random sampling of genes during transmission to subsequent generations. Usually plant species are fractured into large numbers of geographic populations that experience low rates of interpopulation migration. When this is the case, we expect different populations to accumulate different selectively favored alleles because on average only 2sn of the populations will have any particular favored mutation fixed (where n is the number of populations). In addition, different geographic populations experience different environments, and consequently favor different adaptations. For these reasons, we expect the various local populations of a plant species to harbor different adaptive gene combinations (Wright, 1969).

Calculations like those above lead us to infer that many neutral polymorphisms should be segregating in moderately large populations at any point in time. Indeed, the probability of drawing two identical alleles at a locus (F), assuming neutrality and assuming an infinite number of possible allelic states, is approximately, $F = 1/(1 + 4N\mu)$ (Crow and Kimura, 1970). If $N > \mu$, F is small and most draws will involve nonidentical alleles at a locus. These considerations lead to an operational definition of a neutral gene as a gene with a selective coefficient, $s < 1/N$ (Kimura and Ohta, 1971). It also follows that levels of heterozygosity ($H = 1\text{-}F$) will be high in random mating populations when F is small.

Some patterns of selection lead to the retention of genetic diversity. These patterns include overdominant selection, frequency dependent selection and multiple niche selection. What fraction of the standing genetic variation is neutral and what fraction is selected when overdominant selection is possible? This question was considered by Crow and Kimura (1970) and they showed that reasonable choices of s, μ and N lead to the conclusion that the majority of polymorphic alleles are expected to be neutral. It is more difficult to obtain a quantitative evaluation of this question for frequency dependent selection or multiple niche selection, however, the fact that new neutral alleles are expected to arise in appreciable numbers in every generation when $N\mu > 1$, makes it likely that most molecular variants will be neutral.

These theoretical arguments constitute the basic foundations of the neutral theory of molecular evolution (Kimura, 1984). This theory posits that most molecular polymorphisms and most nucleotide substitutions in the course of molecular evolution are selectively neutral. Moreover, it is for all practical purposes impossible to distinguish weakly (or even moderately) selected genes from neutral genes in direct experiments to measure selective coefficients (Lewontin, 1974). For these reasons we must approach the molecular assessment of genetic diversity from the perspective that most of the variants are selectively equivalent. A corollary of this proposition is that most molecular variants have no present or future value for the phenotypic improvement of crop plants.

Geographic patterns of genetic differentiation are more difficult to assess. We have already seen that there are good theoretical grounds for expecting the fixation of different adaptive genes in different geographic populations. The phenotypic differentiation observed among geographic populations may often reflect the fixation of adaptive genes and gene combinations. However, we also expect the random fixation of neutral variants among weakly connected geographic populations and there are no good *a priori* methods for distinguishing adaptive from neutral variants. As a consequence, the mere correlation between molecular variants and adaptive phenotypes when taken over geographic populations is insufficient to establish a causal relationship.

THE PHENOTYPE VERSUS GENOTYPE DILEMMA

The distinction between phenotype and genotype was first drawn by the Danish geneticist W. Johannsen shortly after the turn of the century. Molecular assays of genetic diversity are at or close to the genotypic end of the spectrum. We usually do not know the functional properties - that is the effects on metabolism, physiology or development - of most DNA sequence changes. As a consequence, it is usually not possible to predict the phenotypic effects of molecularly defined allelic alternatives. At one level the neutrality hypothesis can be interpreted as claiming that most DNA sequence variants (e.g. nucleotide differences) code for functionally equivalent and hence phenotypically identical isoalleles.

A host of phenotypic characters have clear adaptive value and are important in plant improvement. These include traits like cold tolerance, drought resistance, growth rate differences, time of reproductive maturity, disease resistance, etc. Some of these phenotypes appear to have a simple genetic basis as is often the case for disease resistance. Others have a complex genetic basis where the action of individual genes can not be resolved (e.g. growth rate, cold tolerance etc). The classical science of quantitative genetics provided a methodological framework for analyzing the genetic determination of complex phenotypes, but it could not identify the individual genes that subsume these phenotypes. Thus the dilemma: we could study molecular polymorphisms, but could not identify their phenotypic consequences, if any, or we could study phenotypes, but could not resolve their individual genetic determinants.

Molecular polymorphisms may provide a way out of this dilemma. The development of high density maps of plant genomes and the use of these markers to identify the chromosomal locus of major factors that determine complex phenotypes has been one of the most promising developments in plant genetics (Tanksley, 1983; Helentjaris et al., 1985; Stuber, 1990).

USES OF MOLECULAR MARKERS

Isozymes and Enviromnental or Phenotypic Correlations

A large number of isozyme surveys of genetic variability in plant species have now accumulated. Although these surveys have revealed large stores of variation, some major patterns are clear. For example, outbreeding species tend to have greater levels of genetic diversity and levels of variation often are correlated with life history patterns (Hamrick and Godt, 1989). A number of studies have uncovered correlations of geographic and ecological variables with allele frequency distributions. For example, early studies of isozyme variation in *Avena barbata* in California revealed striking correlations of allele and multilocus genotype frequencies with major climatic regions (Clegg and Allard, 1972) and these correlations extended down to micro-geographic regions as well (Hamrick and Allard, 1972; Allard et al, 1972). *A. barbata* is believed to have been introduced from Spain to California during the Spanish mission period approximately 400 years ago. Recent work on the distribution of isozyme variation in Spain supports the notion that allelic diversity in *A. barbata* in California is a subset of that found in Spain (Garica et al., 1989). However, the strong environmental correlations with multilocus genotype seen in California are not replicated in Spain (Perez de la Vega, et al., 1991). Apparently, these environmental associations have arisen since the introduction of *A. barbata* to California. These data indicate that spatial patterns of genetic variation can be established relatively rapidly.

Nevo (1987) has emphasized the correlation of isozyme patterns with ecological and phenotypic variables and he has argued that isozyme markers can be used as predictive markers for important quantitative traits that are of value in crop improvement. Surveys of isozyme diversity in wild emmer wheat by Nevo and associates have employed univariate and multivariate statistical analyses to search for correlations between allozyme frequencies, seed protein content (e.g. Nevo et al., 1986), pathogen resistance (e.g. Nevo et al., 1985), and ecological variables (e.g. Nevo and Beiles, 1989). These studies usually reveal correlations between allele frequencies and other phenotypic variables. According to the population genetic theory reviewed above, such correlations are expected to appear owing to the random differentiation of geographically separated local populations. In general, these correlations can not be assumed to indicate causal relationships between allele frequency variables and other phenotypic variables. Linkage disequilibria among genes at different loci can arise from a covariance in gene frequencies over local populations (Nei and Li, 1973). Intercrossing of individuals from separate populations should lead to the rapid dissolution of these correlations. As a consequence, allele frequency variables can not, in general, predict the transmission of other valuable phenotypic traits beyond the initial generation of intercrossing. On the other hand, ecological factors can provide an important guide to useful genotypes has been emphasized by Rick (1991) in the case of tomato *(Lycopersicon esculentum* L.).

The Geographic Structure of Crop Plant Diversity

It is obvious that careful geographic surveys of genetic diversity are important adjuncts to the effective collection of plant genetic resources. Thorough surveys, carefully analyzed, may provide clues as to centers of origin of domesticated plants. Several workers have attempted to use cluster analysis of genetic distances to define the relationships among closely related cultivated species. Loaiza-Figueroa et al. (1989) characterized isozyme diversity among 186 accessions of domesticated, semidomesticated and wild forms of pepper *(Capsicum).* Based on these analyses they postulated a primary center of domestication for cultivated *C. annuum* in eastern Mexico. A possible second center of domestication was also postulated in the state of Nayarit in western Mexico.

Studies of seed protein variation and isozyme variation in the common bean *(Phaseolus vulgaris)* by Gepts (1991; 1990) have established two independent centers of domestication for this important crop plant: one center in Mexico and the other in the southern Andes. Isozyme data from beans *(Phaseolus vulgaris)* also reveal an interesting contradiction between morphological and biochemical variation patterns. Beans appear to have become more diverse morphologically during and after domestication, while at the same time biochemical diversity declined. Gepts (1991) suggested that human domesticators selected for morphological diversity, but biochemical diversity was reduced owing to reduced population size (genetic drift) and selection. Selection has the general effect of reducing the number of lineages in the population or species and consequently acts to reduce the level of standing genetic diversity.

Maize is an exceptionally variable crop species. There are some 300 races of maize (Goodman, 1990) and maize is among the most variable annual plant species known for isozyme variants (Doebley et al., 1985). Isozyme analysis of 34 Mexican and 19 Guatemalan races of ;maize have revealed broad allele frequency associations with altitude and geographic region (Doebely et al., 1985; Bretting et al., 1990). Analyses of genetic distance data agree in resolving separate lowland and highland phylogenetic lineages. Considerable genetic diversity is also evident within these broad groupings that presumably reflects local patterns of selection by indigenous cultivators. These kinds of analyses provide very useful guides for the efficient collection and preservation of maize germplasm.

Chloroplast DNA in Surveys of Crop Plant Diversity

Early studies of restriction site variation in plant species concentrated on the chloroplast genome (cpDNA). This molecule is relatively small [averages about 150,000 base pairs (bp)] and it was a primary object of molecular study during the late 1970s and the early 1980s, owing both to its small size and central importance in photosynthesis. The result was a rich accumulation of molecular information that provided a base for biosystematic (Clegg and Durbin, 1990) and evolutionary study (Clegg et al., 1991). Early in the evolutionary study of cpDNA it became evident that this molecule evolved at a conservative rate (Curtis and Clegg, 1984; Zurawski and Clegg, 1987). A consequence of this slow rate of evolution is a low level of intraspecific polymorphism. Despite depauperate levels of variation, a number of investigators have attempted to characterize polymorphism within crop plant species. A summary of these investigations is contained in Clegg et al. (1991). We will discuss two cpDNA studies to illustrate the limitations of cpDNA as a tool in germplasm characterization.

Relative to isozyme surveys, studies of restriction site variation have a major limitation: the number of accessions that can be surveyed is small. This is a consequence of the much more labor intensive DNA preparation and Southern transfer technologies. Sample sizes in studies of cpDNA diversity reflect this limitation. For example, in an early study of cpDNA diversity within wild and cultivated barley *(Hordeum vulgare* and *H. spontaneum)* Clegg et al., (1984) surveyed a sample of just 20 lines (11 lines of *H. spontaneum* and 9 lines of *H. vulgare).* Despite the very small sample size this study did suggest a reduction in cpDNA diversity from wild to cultivated forms. These workers also investigated the correlation between isozyme and cpDNA diversity and concluded that the two were poorly correlated in the sample of 20 lines. Interestingly, one common cpDNA restriction pattern in wild barley was the exclusive type found among a sample of four modern cultivars. Subsequent surveys of cpDNA diversity in barley with sample sizes of around 100 have largely confirmed these early results (Holwerda et al., 1986; Nealeetal., 1988).

An early cpDNA survey of 12 pearl millet *Pennisetum glaucum* cultivars was even less informative, revealing no intraspecific polymorphism (Clegg et al., 1984). Subsequently, a larger survey was conducted that included 25 wild and 54 cultivated accessions of pearl millet sampled from throughout its range, but this larger effort still failed to uncover any intraspecific cpDNA polymorphism (Gepts and Clegg, 1989). These, and the results of numerous other cpDNA surveys of intraspecific polymorphism (reviewed in Clegg et al., 1991), lead to the conclusion that surveys of cpDNA polymorphism probably have little utility for assessing crop plant diversity.

RFLPs and Crop Plant Diversity

The earliest RFLP studies of genetic diversity in crop plants and their wild relatives concentrated on the ribosomal DNA (rDNA) encoding regions. The rDNA is organized into one or more chromosomal blocks of tandemly repeated 18S, 5.8S and 26S sequences (Dvorak, 1990). Typically there are more than 1000 and sometimes up to 10,000 copies of the rDNA repeating unit per plant genome (Long and Dwaid, 1980). The 18S and 26S sequences are very highly conserved in evolution and usually exhibit little intraspecific polymorphism. Part of the tandem repeat unit, known as the external or nontranscribed spacer region (NTS), evolves more rapidly and is associated with moderate levels of intraspecific polymorphism. Most of the polymorphism in the NTS region is the result of length variation among copies of a subrepeat sequence that varies from about 100 bp (in grasses) to more than 300 bp (in peas) and is tandemly repeated in the NTS region. The primary cause of length variation is thought to be unequal recombination among NTS subrepeat units (Appels and Dvorak, 1982; Dvorak, 1990).

Molecular studies of rDNA polymorphism were among the first direct DNA based surveys of plant molecular diversity because (1) it was the first plant gene to be cloned; (2) the highly abundant rDNA made Southern hybridization particularly easy; (3) the highly conserved nature of the 18S and 26S sequences meant that DNA-DNA hybridization could be carried out using a heterologous probe against a wide taxonomic range of species; and (4) reasonable levels of intraspecific polymorphism were found to be associated with the NTS region. Careful studies of rDNA polymorphism in barley by Allard and associates (Allard, 1988; Sagahi-Maroof et al., 1984) revealed two rDNA loci (denoted *Rrnl* and *Rrn2)* with a total of twenty alleles among them. All alleles are length polymorphisms and represent a series of eight alleles at *Rrnl* and 12 alleles at *Rrn2* that form a regular ladder of subrepeat copy number increases.

Detailed surveys of rDNA variation within wild and cultivated barleys, based on a sample of 359 accessions (267 wild and 92 cultivated), have been reported by SaghaiMaroof et al. (1990). While all populations of wild barleys were polymorphic, a single length variant at *Rrnl* and *Rrn2* was most frequent across the entire collection of wild barleys. Most of the twenty length variants at *Rrnl* and *Rrn2* were infrequent or rare. It is noteworthy that the length variant most common among the cultivated accessions at *Rrn2* was rare in the sample of wild barleys.

A geographic survey of rDNA variation among wild and cultivated forms of pearl millet revealed higher levels of polymorphism among wild as compared to cultivated forms, which were nearly monomorphic for a common rDNA pattern (Gepts and Clegg, 1989). Interestingly, RFLP polymorphism associated with the Adh1 locus of pearl millet did not show a reduction in variability between wild and cultivated forms (Gepts and Clegg, 1989). A complicating feature of this and similar studies was the existence of rDNA heterogeneity within individual genomes. Evidently some tandem repeat units had one mutational form and others had a second or third form. This quantitative variation in allele type within genomes complicates the scoring of rDNA variants on a population basis [estimates of allele copy number per genome have been obtained for barley using slot-blot assays (Zhang et al., 1990)]. The somewhat discordant results for the rDNA versus *Adhl* RFLPs are not surprising since each represents one *(Adhl)* or a small number of chromosomal loci (rDNA). Accurate assessments of patterns of genetic diversity for gene conservation should rest on averages taken over a large number of loci.

One area where the combination of chloroplast and nuclear RFLPs has been particularly fruitful is in uncovering the role of hybridization in crop plant origins. The cpDNA is typically transmitted uniparentally and usually maternally in angiosperms (see Clegg et al., 1991 for a discussion of transmission patterns). Uniparental (maternal) transmission provides a basis for separating the contribution of maternal lineages from paternal lineages when hybridization is suspected. Furnier et al. (1990) used this approach to investigate the origin of one of the major cultivated varieties of avocado *(Persea americana* var. *guatemalensis).* RFLPs based on a cellulase gene family, rDNA and cpDNA were used to survey 19 cultivars chosen from each of the three botanical varieties (var. *americana,* var. *dymifolia,* var. *guatemalensis)* and four other closely species of *Persea.* The RFLP data suggest that var. *guatemalensis* is a hybrid between *P. nubigena* and *P. steyermarkii.* The data also suggest that the three varieties of cultivated avocado were domesticated independently. Clearly, this kind of information is of great utility in defining the gene pool of cultivated avocado, an essential prerequisite to germplam collection.

During the past five years a large number of anonymous clones of random DNA fragments have been employed as RFLPs in genome mapping efforts. Because these RFLP loci are randomly selected, the coding function, if any, of the cloned DNA is unknown. The primary criterion for inclusion of a particular clone in a mapping program is pre-existing polymorphism. At first sight these anonymous clones might seem to be poor choices for surveys of molecular diversity because nothing is known of their functional significance. On reflection, it is clear that this concern is analogous to concerns about the value of survey information based on "neutral" isozyme variants. We have already seen that isozyme surveys provide very useful information for gene conservation whether neutral or not. Similarly, surveys with random RFLP markers should be very useful in defining genetic relationships among geographic collections. A considerable advantage of RFLP markers is the virtually unlimited number of polymorphic loci available to the investigator. This permits the calculation of genetic distance or similarity statistics based on very good averages across loci. Such data should be quite helpful in revealing genetically divergent isolates that might harbor unusual genetic variants.

In one of the few surveys based on anonymous RFLP marker loci, Miller and Tanksley (1990) used 40 clones, whose chromosomal locations had been previously mapped, to characterize genetic diversity in *Lycopersicon*. The sample included 156 plants from eight species and cluster analyses of the resulting genetic distance statistics were consistent with the current classification of the species. However, one mountain accession of *L. peruvianum* var *humifusum* was found to be sufficiently distinct to possibly warrant classification as a separate species. Because a large number of loci were surveyed, Miller and Tanksley (1990) were able to estimate the likelihood that a given accession would add new alleles to the collection. This likelihood was two-fold higher for the more polymorphic self-incompatible species than for the self-compatible species. This means that twice the collecting effort will be necessary for the self-compatible forms to include an equivalent sample of genetic diversity. Even more striking is the estimate that a single accession of *L. peruvianum is* twenty times more likely to add new alleles than is a single accession of cultivated tomato.

A recent innovation on the RFLP approach (called the RAPD method), that is based on the use of random eight to ten bp synthetic DNA primers coupled with the polymerase chain reaction (PCR) technique, promises to greatly increase the efficiency of RFLP surveys (Williams et al., 1990). The method rests on the random amplification of arbitrary DNA segments (in the range of 200bp to 2000 bp in size) that are bounded by a primer sequence. For any given primer, the number of segments that can be amplified is expected to be small and empirical work appears to support this expectation. The great appeal of the method is that the time consuming steps of Southern transfer and hybridization are eliminated because the RFLP patterns can be visualized in an agarose gel. This means that much larger sample sizes are feasible than is the case for standard RFLP work. A draw back of the method is that DNA fragments corresponding to several chromosomal loci are revealed in any given amplification and the assignment of alleles to loci is not always evident without genetic analysis. In addition, polymorphic bands tend to exhibit a presence/absence pattern of variation that leads to recessive/dominant phenotypic classifications. A consequence of the presence/absence pattern is that allele frequencies can not be estimated directly without the use of progeny testing or subsidiary hypotheses like random mating. Despite these limitations, this method is likely to be widely employed in the future. It should be especially useful for establishing a "DNA fingerprint" for accessions or cultivars. Such an approach would help in gene bank management in the identification of redundant collections.

As alluded to earlier, high density maps of plant genomes have opened up the possibility of analyzing the genetic basis of major phenotypic traits associated with plant domestication. In the first such study, Doebley et al. (1990) employed RFLP markers to dissect the genetic determination of the major traits that distinguish maize *(Zea mays* L. *ssp. mays)* and teosinte *(Zea mays ssp. mexicana)*. The resulting analysis of a maize-teosinte F2 population for nine of the principle traits distinguishing maize from teosinte showed that each trait was under multigenic control. The marker data also showed that genes on chromosome 1 accounted for much of the variation associated with seven of the traits. Chromosomes 2, 3 and 4 were associated with major genes affecting three of the traits. This kind of complex analysis is difficult and demanding of time and resources. Nevertheless, it provides a much finer resolution of the nature of genetic changes associated with crop plant domestication than had previously been available. It is obvious that map information of this kind could play an important role in devising conservation strategies.

Allelic Diversity at The Complete DNA Sequence Level

The ultimate descriptor of genetic diversity is the DNA sequence of a sample of genes. What kind of information can be obtained from DNA sequence data that would justify the considerable effort required to obtain the data? A sample of gene sequences can provide important information on the history of the population and possibly on past selective events at the gene of interest. The theoretical basis for inferring genealogical histories is contained in coalescence models (Hudson, 1991). These models ask when in the past a sample of alleles (separate lineages) coalesce into a single ancestral lineage (Fig. 1). The expected time of the coalescence event is a function of the mutation rate, the effective population size and selection parameters. Coalescence theory for neutral genes tells us that the average time back to the coalescence of all sample lineages into a single ancestral lineage (T) quickly approaches 4N (as the number of gene sequences in the sample increases). The standard deviation of T approaches 2.1N, so we expect considerable stochastic error from locus to locus in estimates of T.

Now consider a gene locus that has been subject to a recent selective sweep, where a favorable mutation has replaced a preexisting allele. We have seen that the expected time to fixation for a selectively favored gene is relatively short. All other polymorphic nucleotide sites within several thousand bases of the favored mutation would also be carried to fixation as the

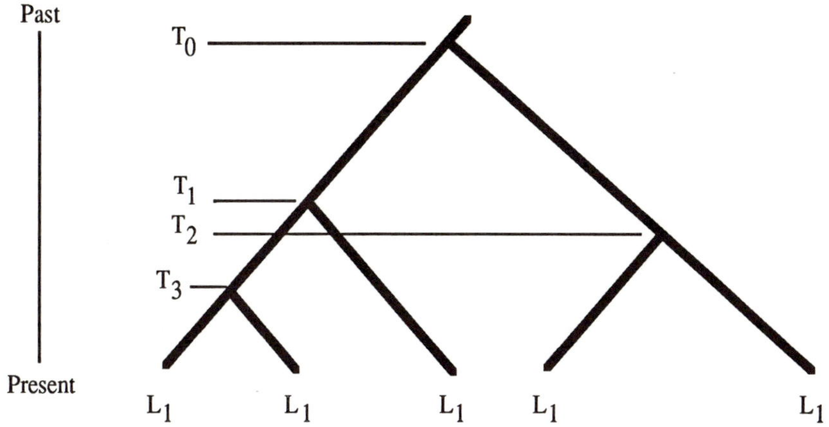

Fig. 1. Illustrates the genealogies of a sample of genes (L_1 to L_5). The Coalescence time of the sample is denoted T_0, and the coalescence times for the individual lineages are T_1, T_2, and T_3.

result of hitchhiking selection. The consequence of a recent selective sweep is a substantial reduction in nucleotide polymorphism in the neighborhood of the selected nucleotide relative to neutral genes. (The neighborhood size depends on the strength of selection and recombination, but is likely to be several thousand base pairs in length.) A very shallow history of sample lineages may reflect a recent selective sweep (illustrated in Fig. 2), and therefore, may indicate alleles that harbor adaptively important mutations. Coalescence theory may also provide a framework for recognizing regions that have been held polymorphic by some form of balancing selection. Kreitman (1990) has presented detailed analyses of a sample of sequences for the alcohol dehydrogenase locus *(Adh)* from *Drosophila melanogaster* that strongly suggest a pattern of balancing selection favoring the retention of the fast and slow isozyme forms at this locus. The basic argument is that balancing selection will lead to the retention of two or more alleles in the population well beyond the expected 4N coalescence time. Other polymorphic nucleotide sites within a few hundred bp of the selected site are also expected to be retained for a time greater than 4N. As a consequence, levels of nucleotide polymorphism are expected to be elevated in the neighborhood of the selected site. The *Adh* data from *D. melanogaster* do indicate a higher level of polymorphism around the presumed site of selection. Evidently detailed molecular studies may aid in identifying the location of adaptively important mutations and may thus help resolve the phenotype/genotype dilemma. Clearly, information of this kind is of great value is gene conservation.

Fig. 2. Illustrates the reduction in lineages associated with a selective sweep. The * marks the occurrence of a new favored mutation in a particular lineage.

What do we know about DNA sequence variation for samples drawn from plant species? The short answer is very little. A recent review of our knowledge of plant molecular diversity was only able to identify four genes where sequence data had been obtained for two or three alleles (Clegg, 1990). In all four cases [alcohol dehydrogenase-l *(Adhl),* the bronze locus, the shrunken locus and phosphoenolpyruvate carboxylase (PEPCase)] the data indicated considerable, and roughly equivalent levels of sequence polymorphism; however, these data were not sufficient for coalescence analyses like those described above. Recently a study of *Adhl* evolution in the grass family has appeared which presents an estimate of approximately 2.6 million years for the coalescence time for the *Adhl-F* and *Adhl-S* alleles of maize (Gaut and Clegg, 1991).

Recent studies of DNA sequence polymorphism at genes encoding selfincompatibility alleles in Solanaceous plant species indicate that some alleles may be as old as 30 million years (Clark et al., 1991). This great age is entirely consistent with the biology of self-incompatibility alleles which are subject to a form of frequency-dependent balancing selection, owing to their effect on the mating system. While it is far too early to make any kind of generalization, these two studies do suggest that some polymorphisms may be older than their constitutient species. We need to be able to increase the data base for *Adhl* and other genes in maize to ask whether the estimates for particular genes are heterogeneous, thus indicating a selective history. The advent of PCR mediated direct sequencing has made it feasible to determine the sequence of a sample of genes and we may expect to see such studies appear in the near future.

Because DNA is a relatively stable molecule, it can be recovered from ancient plant materials and particular genes can be PCR amplified and sequenced. This approach has been employed to study DNA sequence diversity for a small region (approximately 300 bp) of the *Adh2* locus of maize in samples from ancient maize remains (ranging from 1,500 to 4,500 years old) and from modern cultivars (Goloubinoff et al., 1991). The total sample included six modern maize alleles, six alleles from ancient maize materials, eight alleles from teosinte and one from *Tripsacum*. Genetic distance and parsimony analyses of the sequence data show that the modern and ancient maize sequences and the teosinte sequences cluster together. Goloubinoff et al. (1991) conclude that teosinte and maize share a single broad gene pool that predates the domestication of maize. The ability to study ancient materials directly allows gene conservationists to assess the loss in diversity associated with the activities of modern civilizations. These kinds of direct assessments may lead to the recognition and rescue of gene variants that were once widespread but are now threatened. The utility of ancient DNA studies, of course, depends on the kinds and quality of ancient plant remains available to the investigator.

CONCLUSIONS

Since their appearance nearly twenty years ago, molecular (or recombinant DNA) techniques have contributed to our knowledge of plant genetic diversity. The application of molecular methods for survey purposes has been expensive in terms of time and resources, and as a consequence, sample sizes have been small by isozyme survey standards. Moreover, the vast bulk of data on molecular diversity comes from a very, small number of domesticated plant taxa, principally maize and tomato. The question to confront is does the quality of the data justify the effort? In attempting to answer this question, it is important to realize that the pace of technological innovation in molecular methods has been remarkable. New survey methods have quickly evolved from the PCR technique (e.g. the RAPD method and direct sequencing of PCR products) and these have greatly accelerated our ability to acquire survey data. Any answer to this question must anticipate continued increases in the efficiency of molecular survey techniques. The principle advantages of molecular data are (1) a very large number of gene loci can be evaluated; (2) molecular markers provide powerful tools for dissecting the genetic determination of important traits; (3) DNA sequence data may provide a means of identifying adaptively important mutations of unknown phenotype; and (4) DNA can be studied from ancient as well as extant materials, thus providing an unparalleled view of the historical gene pool of a plant species. These advantages are so fundamental that we can safely conclude that molecular methods have an important role to play in the assessment and conservation of plant genetic resources.

ACKNOWLEDGMENT

Supported in part by NIH grant GM 45144.

REFERENCES

Allard, R.W., 1989, Genetic changes associated with the evolution of adaptedness in cultivated plants and their wild progenitors, *J. Heredity* 79:225-238.

Allard, R.W., Babbel, G., Clegg, M.T., and Kahler, A.L., 1972, Evidence for coadaptation in *Avena barbata, Proc. Natl Acad. Sci. USA* 69: 3043-3048.

Appels, R., and Dvorak, J., 1982, Relative rates of divergence of spacer and gene sequences within the rDNA region of species of the Triticeae: Implications for the maintenance of homogeneity of a repeated gene family, *Theor. Appl. Genet.* 63:361-365.

Bretting, P.K., Goodman, M.M., and Stuber, C.W., 1990, Isozymatic variation in Guatemalan races of maize, *Arner. J. Bot.* 77:21 1-225.

Brown, A.H.D., and Clegg, M.T., 1983, Isozyme assessment of plant genetic resources, *in:* "Isozymes: Current Topics in Biological and Medical Research Volume 11: Medical and Other Applications," M.C. Rattazzi, J.C. Scandalios, and G.S. Whitt, eds., Alan R. Liss, New York. pp 285-295.

Brown, A.H.D., Frankel, O.H., Marshall, D.R., and Williams, J.T., (eds.), 1989, "The Use of Plant Genetic Resources," Cambridge University Press, Cambridge, pp 382.

Clark, A. G., and Kao, T.-H., 1991, Excess nonsynonymous substitution at shared polymorphic sites among self-incompatibility alleles of Solanaceae, *Proc. Natl. Acad. Sci. USA,* (in press).

Clegg, M.T., 1990, Molecular diversity in plant populations, *in:* "Plant Population Genetics, Breeding and Genetic Resources," A.H.D. Brown, M.T. Clegg, A.L. Kahler and B.S. Weir, eds., Sinauer Associates, Sunderland, MA. pp 98-115.

Clegg, M.T., and Allard, R.W., 1972, Patterns of genetic variation in the slender wild oat, *Avena barbata, Proc. Natl. Acad. Sci. USA* 69:1820-1824.

Clegg, M.T., and Durbin, M.L., 1990, Molecular approaches to the study of plant biosystenatics, *Aust. Syst. Bot.* 3:1-8.

Clegg, M.T., Learn, G.L., and Golenberg, E.M., 1991, Molecular evolution of chloroplast DNA, *in:* "Evolution at the Molecular Level," R.K. Selander, A.G. Clark, and T.S. Wittham, eds., Sinauer Associates, Sunderland, MA. pp 135-149.

Clegg, M.T., Rawson, J.R.Y., and Thomas, K., 1984, Chloroplast DNA variation in pearl millet and related species, *Genetics* 106:449-461.

Clegg, M.T., Brown, A.H.D., and Whitfeld, P.R., 1984, Chloroplast DNA diversity in wild and cultivated barley:implications for genetic conservation, *Genet. Res.* 43:339-343.

Cohen, J.L., Williams, J.T., Plucknett, D.L., and Shands, H., 1991, Ex situ conservation of plant genetic resources: Global development and environmental concerns, *Science* 253:866-872.

Curtis, S.E., and Clegg, M.T., 1984, Molecular evolution of chloroplast DNA sequences, *Molec. Biol. Evol.* 1:291-301.

Crow, J.F., and Kimura, M., 1970, "An Introduction to Population Genetics Theory," Harper and Row, New York, pp 591.

Day, P.R., Allard, R.W., Alvim, P. de T., Ashri, A., Barton, J.H., Buttel, F.H., Chang, T.-T., Creech, J.L., Dietz, S.M., Evenson, R.E., Fitzhugh, H.A., Goodman, M.M., Hardon, J.J., Johnson, V.A., Marshall, D.R., Maunder, A.B., Qualset, C.O., Paroda, R.S., Sastrapradja, S., Smith, C., Spence, J.A., Timothy, D.H., Wilkes, H.G., and Withers, L.A., 1991, "Managing Global Genetic Resources: The US National Plant Germplasm System," National Academy Press, Washington, DC.

Doebley, J., Stec, A., Wendel, J., and Edwards, M., 1990, Genetic and morphological analysis of a rnaize-teosinte F2 population: Implications for the origin of maize, *Proc. Natl. Acad. Sci. USA* 87:9888-9892.

Doebley, J.F., Goodman, M.M., and Stuber, C.W., 1985, Isoyme variation in the races of maize from Mexico, *Arner. J. Bot.* 72:629-639.

Dvorak, J., 1990, Evolution of multigene families: the ribosomal RNA loci of wheat and related species, *in:* "Plant Population Genetics, Breeding and Genetic Resources," A.H.D. Brown, M.T. Clegg, A.L. Kahler, and B.S. Weir, eds., Sinauer Associates, Sunderland, MA. pp 83-97.

Ehrlich, P.R. and Wilson, E.O., 1991, Biodiversity studies: science policy, *Science* 253:758-762.

Frankel, O.H., and Bennett, E., 1970, Genetic resources, in: "Genetic Resources in Plant: Their Exploitation and Conservation," O.H. Frankel, and E. Bennett, eds., Cambridge University Press, Cambridge. pp 7-17.

Furnier, G.R., Cummings, M.P., and Clegg, M.T., 1990, Evolution of the avocados as revealed by DNA restriction fragment variation, *J. Heredity* 81:183-188.

Garcia, P., Vences, F.J., Perez de la Vega, M., and Allard, R.W., 1989, Allelic and genotypic composition of ancestral Spanish and colonial Californian gene pools of *Avena barbata:* Evolutionary implications, *Genetics* 122:687-694.

Gaut, B.S., and Clegg, M.T., 1991, Molecular evolution of alcohol dehydrogenase 1 in members of the grass family, *Proc. Natl. Acad. Sci. USA* 88:2060-2064.

Gepts, P., 1990, Biochemical evidence bearing on the domestication of *Phaseolus* (Fabaceae) beans, *Econ. Bot.* 44:28-38.

Gepts, P., 1991, Biotechnology sheds light on bean domestication in Latin America, *Diversity* 7:49-50.

Gepts, P., and Clegg, M.T., 1989, Genetic diversity in pearl millet *(Pennisetum glaucum* [L.] R. Br.) at the DNA sequence level, *J. Heredity* 80:203-208.

Goodman, M.M., 1990, Genetic and germplasm stocks worth conserving, *J. Heredity* 81:11- 16. Hamrick J.L., and Allard, R.W., 1972, Microgeographical variation in alloyme frequencies in *Avena barbata, Proc. Natl. Acad. Sd. USA*. 69:2000-2004.

Hamrick, J.L. and Godt, 1990, Allozyme diversity in plant species, *in:* "Plant Population Genetics, Breeding and Genetic Resources," A.H.D. Brown, M.T. Clegg, A.L. Kahler, and B.S. Weir, eds., Sinauer Associates, Sunderland, MA, pp 43-o3.

Helentjaris, T., King, G., Slocum, G., Siedenstang, C., and Wegman, S., 1985, Restriction fragment polymorphism as probes for plant diversity and their development as tools for applied plant breeding, *PlantUol. Biol.* 5:109-118.

Holwerda, B.C., Jana, S., and Crosby, W.L., 1986, Chloroplast and mitochondrial DNA variation in *Hordeum vulgare* and *Hordeum spontaneum, Genetics* 114:1271-1291.

Hudson, R.R., 1991, Gene genealogies and the coalescent process, *Oxford Surveys Evol. Biol.* 7:1-44.

Kimura, M., 1983, "The Neutral Theory of Molecular Evolution," Cambridge University Press, Cambridge.

Kimura, M., and Ohta, T., 1971, "Theoretical Aspects of Population Genetics," Princeton University Press, NJ.

Krietman, M., 1991, Detecting selection at the DNA level, *in:* "Evolution at the Molecular Level," R.K. Selander, A.G. Clark, and T.S. Wittham, eds., Sinauer Associates, Sunderland, MA, pp 204-221.

Lewontin, R.C., 1974, "Genetic Basis of Evolutionary Change," Columbia University Press, New York. pp 346.

Loaiza-Figueroa, F., Ritland, K., Cancino, J.A.L., and Tanksley, S.D., 1989, Patterns of genetic variation of the genus *Capsicum* (Solanaceae) in Mexico, *Pl. Syst. Evol.* 165:159-188.

Long, E.O., and Dawid, I.B., 1980, Repeated genes in eucaroytes, *Ann. Rev. Biochem.* 49:727-765.

Miller, J.C., and Tanksley, S.D., 1990, RFLP analysis of phylogenetic relationships and genetic variation in the genus *Lycopersicon, 7heor. Appl. Genet.* 80:437-448.

Neale, D.B., Saghai-Maroof, M.A., Allard, R.W., Zang, Q., and Jorgensen, R.A., 1988, Chloroplast DNA diversity in populations of wild and cultivated barley, *Genetics* 120:1105-1110.

Nei, M., and Li, W.-H., 1972, Linkage disequilibrium in subdivided populations, *Genetics* 75:213-219.

Nevo, E., 1987, Plant genetic resources: Prediction by isoyme markers and ecology, *in:* "Isozymes: Current Topics in Biological and Medical Research, Volume 16:Agriculture, Physiology and Medicine," M.C. Rattazzi, J.C. Scandalios, and G.S. Whitt, eds., Alan R. Liss, New York, pp 247267.

Nevo, E., and Beiles, A., 1989, Genetic diversity of wild emmer wheat in Israel and Turkey, *Theor. Appl. Genet.* 77:421-455.

Nevo, E., Grama, A., Beiles, A., and Golenberg, E.M., 1986, Resources of high-protein genotypes in wild wheat, *Triticum dicoccoides* in Israel: Predictive method by ecology and allozyme markers, *Genetica* 68:215-227.

Nevo, E., Moseman, J.G., Beiles, A., and Zohary, D., 1985, Patterns of resistance of Israeli wild emmer wheat to pathogens, I. Predictive method by ecology and allozyme genotypes for powdery mildew and leaf rust, *Genetica* 67:209-222.

Perez de la Vega, M., Garcia, P., and Allard, R.W., 1991, Multilocus genetic structure of ancestral Spanish and colonial Californian populations of *Avena barbata, Proc. Natl. Acad. Sci. USA* 88:12021206.

Rick, C.M., 1991, Tomato resources of South America reveal many genetic treasurers, *Diversity* 7:54-56.

Saghai-Maroof, M.A., Soliman, K.M., Jorgensen, R.A., and Allard, R.W., 1984, Ribosomal DNA spacer-length polymorphism in barley: Mendelian inheritance, chromosomal location, and population dynamics, *Proc. Natl. Acad. Sci. USA* 81:8014-8018.

Saghai-Maroof, M.A., Allard, R.W., and Zhang, Q., 1990, Genetic diversity and ecogeographical differentiation among ribosomal DNA alleles in wild and cultivated barley, *Proc. Natl. Acad. Sci. USA* 87:8486-8490.

Soule, M.E., 1991, Conservation: Tactics for a constant crisis, *Science* 253:744-750.

Stuber, C., 1990, Molecular markers in the manipulation of quantitative characters, *in:* "Plant Population Genetics, Breeding and Genetic Resources," A.H.D. Brown, M.T. Clegg, A.L. Kahler, and B.S. Weir, eds., Sinauer Associates, Sunderland MA, pp 334-350.

Tanksley, S.D., 1983, Molecular markers in plant breeding, *Plant Molec. Biol. Rpt.* 1:3-8.

Williams, J.G.K., Kubelik, A.R., Livak, K.J., Antoni, J., and Tingey, S.V., 1990, DNA polymorphisms amplified by arbitrary primers are useful as genetic markers, *NucleicAcids Res.* 18:6531-6535.

Wright, S., 1969, "Theory of Gene Frequencies," University of Chicago Press. Chicago, Ill.

Zhang, Q., Saghai-Maroof, M.A., and Allard, R.W., 1990, Effects on adaptedness of variations in ribosomal DNA copy number in populations of wild barley *(Hordeum vulgare ssp. spontaneum), Proc. Acad. Natl. Sci. USA* 87:8741-8745.

Zurawski, G., and Clegg, M.T., 1987, Evolution of higher-plant DNA-encoded genes: Implications for structure, function and phylogenetic studies, *Annu. Rev. Plant Physiol.* 38:391-418.

EXPLOITING THE DIPLOID 'D' GENOME CHROMATIN FOR WHEAT IMPROVEMENT

E.S. Lagudah, R.Appels, D. McNeil, and D.P. Schachtman

CSIRO, Division of Plant Industry
GPO Box 1600
Canberra, ACT, 26()1
Australia

INTRODUCTION

Bread wheat *(Triticum aestivum L.)* constitutes a polyploid species with 42 chromosomes derived from three genomes designated A, B and D; they each contribute seven pairs of chromosomes to the total genome. Bread wheat originated from the natural synthesis of a polyploid formed between tetraploid wheat, *Triticum turgidum* (AABB) and the diploid species *Triticum tauschii* (DD; syn. *Aegilops squarrosa* L.) (Kihara 1944; McFadden and Sears 1944; 1946). The D set of chromosomes in the genomic composition of bread wheat are considered relatively recent additions and are less diverged from their putative donor than the A and B genomes. Although the A, B and D genomes are distinct, a varying degree of genetic similarity occurs between them. Relationships between each genomic set of the genetically equivalent chromosomes constitutes homoeologous rather than homologous groups. Within a given homoeologous set, such as the group 1 chromosomes, a nullisomic condition for chromosome lD is compensated for by either an additional pair of chromosome lA or lB. The development of a complete set of 42 so-called nullitetrasomic combinations were used to demonstrate chromosome homoeology in the hexaploid wheat genome from the variety, Chinese Spring (Sears 1966). Homoeologous relationships between wheat chromosomes and several related species of the Triticeae (a tribe within the Poaceae) have been confirmed through cytogenetic and biochemical analysis of alien chromosome addition or substitution lines into wheat. Several wild relatives of wheat have been exploited as sources of new genetic material for the improvement of bread wheat. In using species other than the direct progenitors of bread wheat, introgression of new genetic material is dependent on induced homoeologous recombination between wheat and the alien chromosome. Techniques employed in inducing homoeologous recombination include the use of the high pairing *mutant phl* (Sears 1981), chromosomal rearrangements through tissue culture and irradiation induced or spontaneous translocations. A common problem encountered with the use of alien species in crop improvement is the linkage drag of undesirable genetic material accompanying the target gene being introgressed. The choice of the technique for inducing homoeologous recombination as well as selection methods detecting flanking genetic material around the desirable agronomic trait are crucial in achieving the ultimate goal of producing an improved cultivar with reduced deleterious alien chromatin.

WHY EXPLOIT *T. TAUSCHII,* THE 'D' GENOME PROGENITOR?

The immediate progenitors of bread wheat, particularly those contributing the A and D genomes are comparatively less difficult to manipulate than other alien species when transferring small segments of their chromatin for wheat improvement. There are several features of T. *tauschii* that make it a favorable species for genetic studies and for wheat improvement.

1. ***T. tauschii* is the most widely geographically distributed of the diploid progenitors.** Whereas the progenitor and closely related species of the A and B genome component of bread wheat are localized in West Asia, the natural distribution of T. *tauschii* spans geographically diverse regions in Southern USSR, Turkey, Iran, Afghanistan, Pakistan and China. Several plant expeditions into these regions have resulted in the preservation of well documented representative collections of the species. Some of these include the Kyoto University Scientific Expedition in 1955 led by H. Kihara (Kihara et al. 1965), the University of Reading Expedition (Halloran 1968), Plant Germplasm Resource Centre- Canada, University of California- Riverside, University of Melbourne and Oregon State University.

The evolutionary origin of bread wheat apparently occurred with a small number of genotypes of T. *tauschii* in combination with tetraploid wheat. Furthermore it appears that *T. tauschii* individuals involved in the natural synthesis of hexaploid wheat originated from a restricted region near the Caspian Sea (Nakai 1979, 1981; Nishikawa 1983; Lagudah and Halloran 1989; Lagudah et al. 1991b). As a result of this mode of origin the level of genetic variation in the D genome component of bread wheat is limited (see review Appels and Lagudah 1990). The narrow genetic variation of the D genome of wheat is well illustrated by the absence of polymorphism in the intergenic spacer region of the ribosomal DNA at the Nucleolar organiser region (Nor) on chromosome 5D. Whereas different spacer length genotypes have been detected at the *Nor B1* and *-B2* loci (B genome locations), the D genome encoded *Nor D3* locus contains only a single spacer length variant (*Nor D3a* allele,1.5kb) in a survey of a world collection of wheat (Appels and Dvorak 1982; May and Appels 1987b, 1988). This observation contrasts with the extensive variation at the *Nor* locus in T. *tauschii* (Fig. 1), where in addition to the prevalent *Nor D3a* variant in T. *aestivum,* nineteen other intergenic spacer length genotypes have been reported (Lagudah et al. 1991b). Some of the variants in *T.tauschii* are due to the gain or loss of a 120bp subrepeat in the central spacer region of the *Nor D3* locus, and constitute allelic variants at a single major locus. An intergenic spacer genotype fragment from T. *tauschii,* identical in size to the *Nor D3a* allele of T. *aestivum* has been shown to be located on a chromosome 5D substitution line of T. *tauschii* into hexaploid wheat. This T. *tauschii* variant was significantly associated with the subspecies *strangulata,* previously suggested as the main line of descent of the D genome chromosomes in wheat. Although *T. tauschii is* widespread geographically, analyses of the population genetic structure using the Nor

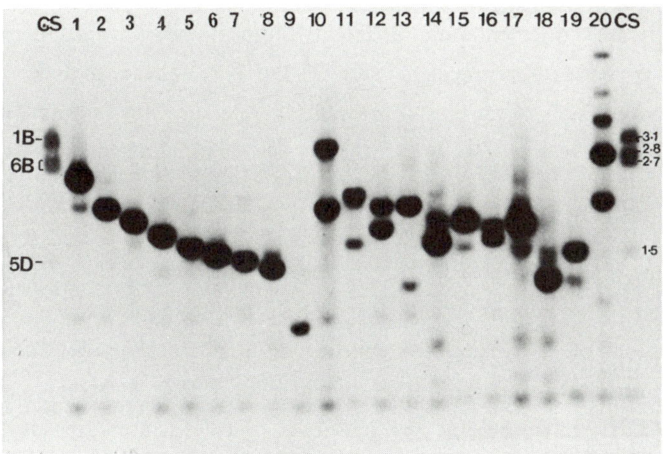

Fig. 1. Autoradiograph of the intergenic spacer region of the Nor locus in T. *tauschii* (1-20) and T. *aestivum, cv.* Chinese Spring (CS). Chromosomal location and the approximate size of the rDNA spacer fragments of Chinese Spring are shown on the left and right CS tracks respectively. Genomic DNA (5ug) from leaf samples were digested with *Taq 1,* electrophoresed, transferred to nitrocellulose membrane and hybridized to the intergenic rDNA spacer clone (pTtDRll); Lagudah et al. 1991b). Lanes 1-20 show the different spacer length genotypes obtained from screening 411 accessions of T. *tauschii* collected within the entire natural geographic distribution.

locus and other biochemical genetic markers has revealed a high diversity in genotypes from Iran and the USSR (Lagudah and Halloran 1989; Lubbers et al. 1991; Lagudah et al. 1991b).

2. **Genetic analysis of *T. tauschii is* easier than that of the allohexaploid bread wheat.** The common occurrence of triplicate loci for several genetic systems in bread wheat requires unequivocal identification of each homoeoallele particularly in genetic linkage studies. The diploid nature of T. *tauschii* and its wide genetic variation has contributed to the rapid development of a molecular genetic linkage map of the D genome (discussed later, Gill et al. 1991; Lagudah et al. 1991a).

3. T. tauschii **genes are readily introgressed into wheat.** The relationship between T. *tauschii* chromosomes and the D genome complement of the hexaploid wheat genome depicts a homologous grouping. Consequently, normal homologous recombination between their respective genomes will facilitate the transfer of new genetic material from the wider gene pool of T. *tauschii* into the D genome of wheat.

4. ***T. tauschii is*** **a proven source of new genes for wheat improvement.** It is well documented from evaluation studies using the wide range of variation collected from different sources that T. *tauschii* can provide new genes for wheat improvement (see Table 1). These include resistance to fungal, nematodes and insect pests as well as tolerance to salinity (Kihara et al. 1965; Kerber and Dyck 1969, 1978; Gill et al. 1986; Appels and Lagudah 1990; Eastwood et al. 1991; Schachtman et al. 1991).

Despite these characteristics, *T. tauschii* has not been fully exploited in thedevelopment of commercial cultivars. A major contributing factor to the under utilization of *T. tauschii* stems from historical perceptions about its agronomic value. In most of the commercial wheat cultivars carrying alien genetic material, resistance to rust diseases have featured prominently. The pioneering work of alien gene transfer into wheat by the late Dr. E. R. Sears, was focussed upon wild relatives showing high resistance to rusts. The few accessions of *T. tauschii* in the possession of Dr. Sears were highly susceptible to all the rust strains tested (Dr. Sears, personal communication), consequently a 'dead- end' was reached in exploiting *T. tauschii* germplasm for wheat improvement. *T. tauschii* accessions spanning their entire natural distribution have since been evaluated for rust resistance using cultures with a wide range of avirulence genes. Incompatible (host resistance) reactions with all three rusts, *Puccinia striiformis, P. graminis* and P. *recondita* were shown to be highly localized geographically. *T. tauschii* accessions predominantly from regions in Iran and southern USSR bordering the Caspian Sea were resistant, whereas all other accessions were compatible (susceptible host) with all three rusts (Mclntosh, Lagudah and Appels, unpublished). This observation clearly highlights the importance of working with representative accessions of T. *tauschii* plant collection, and may account for the absence of resistance in the accessions Dr. Sears had in his collection. Another limiting

Table 1. Fungal, nematode and insect pest resistance derived from *Tritcum tauschii*

Resistance trait	Reference
Fungal diseases	
Rusts *(Puccinia sp.)*	Kihara et al.(1965), Kerber and Dyck(1969), Rowland and Kerber(1974), Kerber (1987), Cox et al. (1990), Lagudah et al. (unpublished data)
Take-all(Gaeumannomyces *graminis)*	Conner et al. (1988)
Karnal bunt *(Neovossia indica)*	Multani et al. (1988)
Septoria nodorum	Loughman et al. (unpublisheddata)
Mycosphaerella grarninicola	May and Lagudah (1992)
Nematodes	
Root knot *nematode(Meloidogyne javanica* and *incognita)*	Roberts et al. (1982), Kaloshian et al (1989)
Cereal cyst nematode(Heterodera *avenae)*	Eastwood et al. (1991)
Insect pests	
Green *bug(Schizaphis graminum)*	Harvey et al. (1980), Joppa (1980)
Hessian *fly(Mayetiola destructor* Say)	Hatchett and Gill (1981), Gill et al. (1985,1986)
Yellow sugar cane *aphid(Siphaflava)*	Merkle and Starks (1985)
Wheat curl *mite(Eriophyes tulipae Keifer)*	Thomas and Conner (1986)
Russian wheat *aphid(Diuraphis noxia)*	Nkongolo *et al.* (1991)

factor in exploiting T. *tauschii,* was based on observations in the performance of artificially synthesized amphidiploids between wheat and other wild relatives. The D genome was assessed to have imparted a lower contribution to the productive capacity of hexaploid wheat when compared to other diploid species (Shebeski 1958).

TRANSFER AND EXPRESSION OF GENES FROM T. *TAUSCHII* INTO HEXAPLOID WHEAT

While several traits of agronomic value have been identified in the diploid D genome, their effectiveness in wheat improvement, must be demonstrated by their expression at the allohexaploid level in the presence of the A and B genome. Transfer of traits from T. *tauschii* to bread wheat can be accomplished by direct crosses into wheat or via artificial synthesis of hexaploid wheat using T. *turgidum* as the AB genomic component.

The direct approach involves crossing T. *tauschii* to bread wheat and rescuing the hybrid (ABDD) embryo on media to overcome postzygotic barriers of embryo-endosperm development. The hybrid plant is normally sterile, but with the occassional formation of restitution nuclei in the egg cells allows it to be backcrossed to bread wheat (male parent). Subsequent backcrosses results in improved fertility and ultimately selection for 42 chromosome plants carrying the introgressed trait from T. *tauschii* . The direction of the cross between T. *tauschii* and bread wheat influences the frequency of hybrid embryos obtained, and the recovery of hybrid plants after embryo culture. A high frequency of seed set usually occurs with T. *tauschii* as the female parent, but the embryo rapidly degenerates due to the formation of a watery endosperm. The timing of embryo rescue is very critical to allow further development and recovery of viable hybrid plants. In crosses with T. *aestivum* as the female parent, seed set tends to be low but a comparatively higher proportion of viable plants are recovered after embryo culture. Direct crosses, where succesful, offers the advantage of reducing the the generation time required to produce advanced breeding lines (Gill and Raup 1987). Breeding lines derived from direct crosses and carrying rust resistance have been field tested in Kansas and in some locations were found to perform better than the control commercial cultivar (Cox et al. 199()). In the direct cross (for transferring traits from *T. tauschii* into wheat) an *a priori* assumption is made that the introgressed trait is sufficiently expressed to allow detection and selection in the background of the recurrent wheat parent.

The alternative approach of indirect transfer of traits from T. *tauschii* into wheat is reliant on the formation of an amphidiploid (synthetic hexaploid) from a cross between tetraploid wheat and T. *tauschii* . Normal development and recovery of hybrid plants is also achieved by embryo culture. The triploid (ABD) hybrids can be induced to form hexaploids using chemicals such as colchicine. Alternatively, spontaneous chromosome doubling occurs with some genotypes of tetraploid wheat, particularly *T. carthlicum (T. persicum)* (Kihara et al. 1965). Once the synthetic hexaploid is formed, the expression of the T. *tauschii* trait can be ascertained. The genomic composition of the synthetic hexaploid is equivalent to bread wheat, consequently, intercrosses between them and subsequent backcrosses into the recurrent bread wheat can then be practised to select for the introgressed trait derived from homologous recombination. Several traits shown in Table 1, for example, resistance to rusts, wheat curl mite, *Mycosphaerella graminicola*, *Herterodera avenae* and salinity have been transfered from T. *tauschii* into wheat using this approach.

The expression of some of the traits listed in Table 1 are modulated by the background of the tetraploid genome. Expression of traits inherited from T. *tauschii* at the allohexaploid level ranges from hypostatic to epistatic interactions (Kerber 1983; Chevre et al. 1989; Eastwood et al. 1991). These interactions have not been considered as recessive or dominant in synthetic hexaploids, since genes controlling traits in the parental tetraploid wheat and T. *tauschii* would interact as homoeologous rather than homologous loci. However some parallels between hypostatic and recessive gene effects and epistasis and dominance are apparent in Table 2. For example, allelic behaviour of winterness in bread wheat is expressed as a recessive trait, and similarly winter habit from T. *tauschii is* suppressed by spring habit from the tetraploid wheat parent. Grain softness is inherited as a dominant trait in bread wheat and in synthetic hexaploids, the hard grain texture of the tetraploid wheat is suppressed by the soft texture derived from T. *tauschii* (Lagudah et al. 1987). In the case of the incompatible rust reaction, resistance in the host is usually dominant whereas in the synthetic hexaploids rust resistance derived from T. *tauschii is* either partially or completely suppressed. The mechanism underlying some of these observations are as yet not fully understood. In one example of intergenomic suppression of traits in the hexaploid wheat genome, the presence of a suppressor gene(s) on chromosome 7D has been identified that inhibits expression of stem rust resistance in the AB genome component (Kerber and Green 1980). The suppressor effect was demonstrated to be also present in chromosome 7D substitution lines of three T. *tauschii* lines tested. Since there exists wide genetic variation in *T. tauschii* it may be possible to identify alternative alleles that lack the suppressor effect, and thus their substitution for the corresponding homoallele in bread wheat may allow expresssion of certain rust resistance genes in the AB genome. In this approach of exploiting the genetic variation in T. *tauschii* for wheat improvement, useful genes present in the tetraploid genome which are otherwise suppressed can be expressed.

Although, suppression of agronomic traits transferred from T. *tauschii* to hexaploid wheat are considered a serious barrier to the transfer of useful traits, it is an advantage in the partial or complete suppression of weedy

Table 2. Transfer and expression of traits from T. *tauschii* (2n=14) to allohexaploid wheat (2n=42)

Trait / Level of expression	Suppression	Partial	Complete
morphology			
brittle rachis		+	
tenacious glumes		+	
winter habit	+		
grain quality			
endosperm protein composition			+
grain softness			+
disease resistance			
Puccinia sp.(rusts)	+	+	
Septoria nodorum	+	+	
Septoria tritici		+	+
Heterodera avenae			+
Greenbug			+
Hessian fly			+
Meloidogyne javanica			+

characteristics such as brittle rachis and tenacious glumes (Table 2). Partial expression of agronomic traits transferred from T. *tauschii* are insome cases, eg. the rusts, still adequate for breeding purposes.

GERMPLASM ENHANCEMENT OF *T. TAUSCHII* DERIVED CHROMATIN

Another factor contributing to the underutilization of T. *tauschii* in wheat improvement is the inadequate germplasm enhancement or prebreeding of the synthetic hexaploids. It is a common practice to synthesize hexaploids, establish the expression level of the transferred agronomic trait, determine chromosomal location and mode of inheritance in most genetic studies without proceeding any further. Underlying the lack of germplasm enhancement is the presumption that the genomic composition of the synthetic hexaploid is homologous to bread wheat and desirable agronomic traits can be readily transferred if required for breeding. However, in most plant breeding programmes, there is a clear preference for elite germplasm, from which numerous combinations of agronomic traits are to be selected for cultivar development. A primary motivation for germplasm enhancement, using T. *tauschii* derived chromatin, is the economic importance of the agronomic trait being exploited. In Australia, the significant reductions in wheat yields due to *Heterodera avenae,* and the limited source of host resistance, has stimulated prebreeding studies in the exploitation of new resistance genes identified in T. *tauschii* (Eastwood et al. 1991). Similarly in North Africa, advanced breeding lines with T. *tauschii* derived Hessian fly resistance are close to cultivar release (B.S.Gill personal communication). Accelerating germplasm enhancement with T. *tauschii* chromatin is also facilitated by the collaborative efforts of geneticists, breeders and agronomists (Cox et al. 1990).

MOLECULAR BASIS FOR GENETIC MANIPULATION

Molecular analysis of genes involved in important developmental processes in eukaryotic genomes is hampered by the absence of well defined gene products controlling these agronomic traits. One avenue for cloning genes where the product is unknown is the use of transposable elements, which alter and tag the target gene being studied, allowing the isolation of the gene using DNA of the transposable element as a probe. In principle this approach can be applied to organisms where transformation with well characterized transposable elements are possible(Baker et al. 1986; Balcells et al. 1991).

An alternative approach which is applicable to organisms where naturally occurring or induced mutants are available is the identification of closely linked DNA sequences to the target trait, which serves as entry points for analysing the target gene region. Molecular genetic markers permit a detailed analysis of plant genomes by allowing the construction of linkage maps, which can provide entry points in studying specific regions of the genome. In recent years, considerable progress has been made in the construction of molecular genetic linkage maps of several plants. Applications of molecular genetic markers and linkage maps in germplasm evaluation and crop improvement has been reviewed by Tanksley et al. (1989). Some of these applications include marker facilitated selection of traits normally difficult to assay, pyramiding of genes with the same phenotype, increased precision in manipulation of chromosome segments and the 'fingerprinting'/identification of plant varieties. In a study of the progeny from two sources of resistance to *Septoria tritici* blotch present in synthetic hexaploids (resistance derived from T. *tauschii)* and a bread wheat cultivar, M1696, segregation ratios observed, revealed the presence of two dominant and independent loci controlling resistance (May and Lagudah 1992). Resistant lines from the progeny resulted from any of the following categories: i) *T. tauschii locus,* ii) M1696 locus, iii) recombinants carrying both loci(i and ii). However, because of the similar phenotype of the resistance, their genotypes could not be fully ascertained. If close linkages between each locus and DNA markers are detected, it will facilitate the genotyping and pyramiding of genes for *Septoria tritici* blotch into a single adapted cultivar.

Some of the major constraints to the rapid development of a molecular genetic linkage map of bread wheat, are the frequent occurrence of triplicate loci due to its polyploid genome relatively low level of restriction fragment length polymorphisms (RFLP) compared to other Triticeae cereals. Furthermore, the varying degree of genetic divergence in the bread wheat genome has resulted in differential levels of polymorphism in their constituent genomes with a ranking order of B>A>D (Dvorak and McGuire 1981; Gale et al. 1990). This differentiation between genomes is reflected in the paucity of polymorphic markers identified in the RFLP-genetic map of the D genome of hexaploid wheat in contrast to the B and A genomes (Liu and Tsunewaki, unpublished).

Genetic Linkage Map of the D Genome

Exploiting the wide genetic variation in T. *tauschii* and its homologous relationship with the D genome of T. *aestivum,* has allowed considerable progress to be made in the construction of the molecular genetic linkage map of the D genome based on T. *tauschii* (Gill et al. 1991; Lagudah et al. 1991a). In constructing the genetic map, divergent T. *tauschii* parental lines were selected based on the frequency of RFLP differences and agronomic traits such as rust resistance. Markers used in the determination of polymorphisms included genomic clones of T. *tauschii* enriched for low copy sequences (using the methylation sensitive restriction endonuclease, *Pst I* digested DNA), cDNA clones from wheat and barley, well defined sequences such as endosperm storage proteins, dehydrins, early methionine polypeptides and isozymes. Computer analysis (using the Mapmaker program, Lander et al. 1987) of the F2 progeny segregation for these markers were used to generate a likelihood map, from three-point and multipoint analysis of suspected linkage groups. The development of the map was assisted by the use of ditelosomic stocks of bread wheat and barley chromosome additions into wheat to confirm the chromosome arm location and homoeology of the biochemical markers (Islam et al. 1981; Hart 1987) (Fig. 2).

Several agronomic traits have been mapped in relation to other markers in the genetic linkage map (Fig. 2). Most of the map positions in Figure 2 were produced from the F2 progeny of the cross AUS18902 (ssp. *eusquarrosa* var. *typica*) and AUS18911 (ssp. *eusquarrosa* var. *meyeri*). On chromosome 1D the storage protein loci important in viscoelastic properties of wheat dough, *Gli-l, Clu-3* (LmwGlu) and *Glu-l,* have been mapped. When taken together with other markers such as the triticin (Tri), early methionine labelled polypeptide (Em) and Adh loci they reveal a conserved gene synteny with the D genome of hexaploid wheat and chromosome lR of rye (Payne 1987; Baum and Appels 1991). Recombination between the *Gli-l* and *Glu-3* loci in the D genome of bread wheat and another T. *tauschii* F2 progeny (AUS 18898 var. *strangulata* x CPI 110730 var. *typica)* have not been detected, which contrasts with the 4% recombination obtained in the T. *tauschii* progeny of AUS 18902 x AUS 18911. Linkage estimates between the centromere and *Glu-l* locus have been reported to vary from 9-33cM; this variation has been attributed partly to the methods employed, namely telocentric and translocation mapping (Payne et al. 1982; Lagudah and Halloran 1988; Singh and Shepherd 1988). However, linkage analysis in different F2 progeny of T. *tauschii* showed the non-randomness of recombination events in different genotypes. For example, the early methionine polypeptide locus (Em) on the long arm of chromosome lD (tightly linked to the centromere) shows a recombination value of 16% with the *Glu-l* locus (18902 x 18911) whereas in the same region in a different cross (18898 x 110730) gives a recombinatiòn frequency of 48%.

The maps of chromosome 3D and 4D shown in Figure 2 are deficient for markers on the short and long arms respectively. A notable feature of some of the regions on the short arm of chromosome 4D, are the cluster of loci involved in germination and early seedling establishment, namely the germination shoot polypeptide (germin), 7S globulin and Adhl and 2 isozymes. The D genome of wheat has been shown to contribute significantly to salinity tolerance, and a major location for the Na+ exclusion trait is located on the long arrn of chromosome 4D (Gorham et al. 1987). Sodium exclusion under saline conditions is considered to be one of the mechanisims contributing to salt tolerance. A partial genetic linkage map produced from the F2 progeny obtained from parental lines with high and low Na+ exclusion (Schachtman et al. 1991) showed the region around the germin locus to account for 20% of the variation in Na+ exclusion using the interval mapping analysis described by Lander and Bostein (1989). Failure to detect significant linkage with the other 80% of the variation, may partly be due to the non coverage in the genetic map of the long arm of chromosome 4D where a major location of Na+ exclusion has been reported.

Some of the other agronomic traits mapped in Figure 2 are the grain softness related protein gene family (Gsp) distal to the *Nor-D3* locus and the dehydration induced proteins *(dhn)* on chromosomes 5D and 6D respectively. Resistance to *Heterodera avenae* (cereal cyst nematode, CCN) from rye and T. *ventricosum* has been located on chromosomes 6R and 6M respectively (Asiedu et al. 1990; Rivoal et al. 1986), and the former is mapped to a region adjacent to the *Got 2* locus (Dundas et al. 1990). In T. *tauschii,* high levels

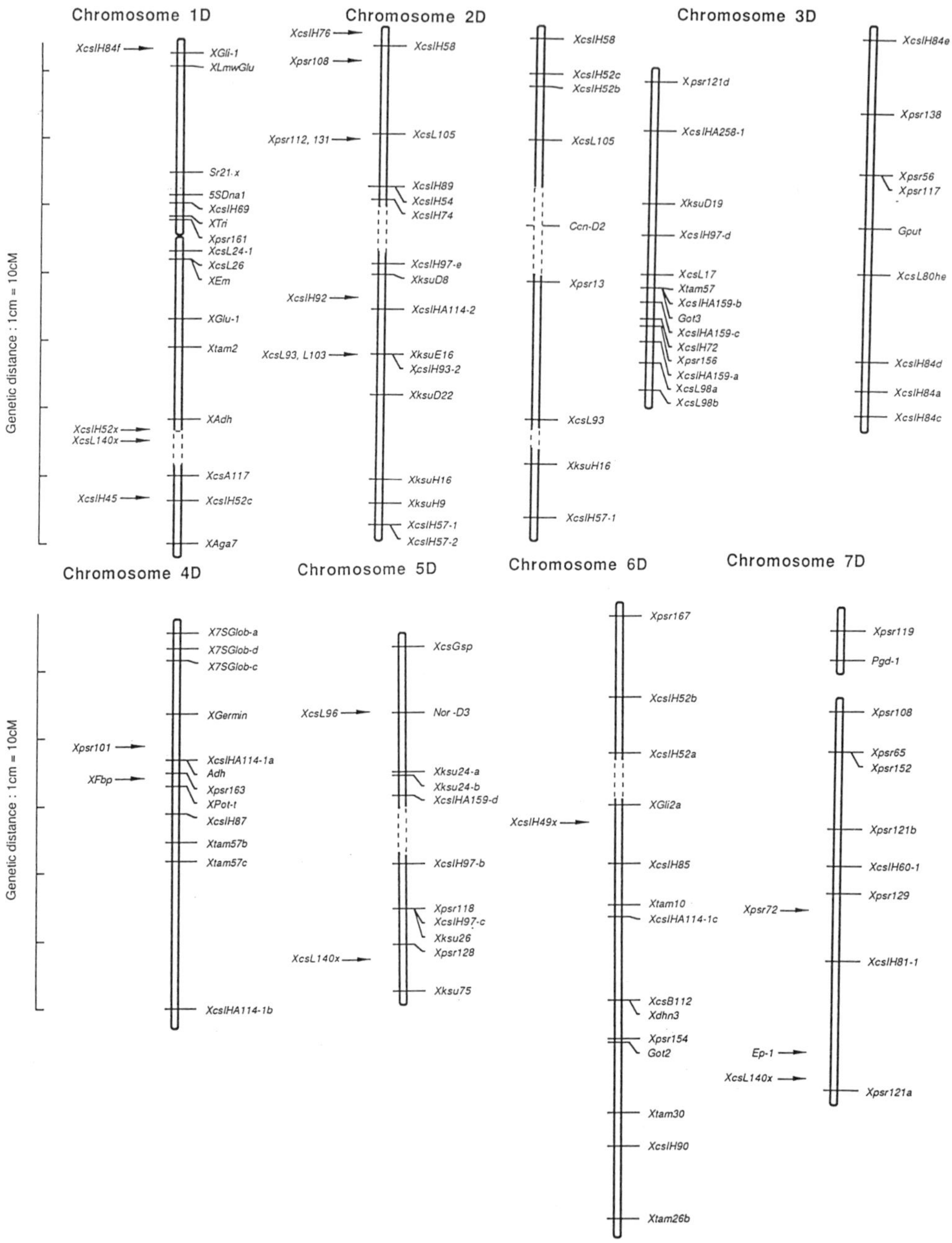

Fig. 2. Genetic maps of chromosomes ID, 2D, 3D, 4D, SD, 6D and 7D produced from the F2 progeny of a T. *tauschii* cross (AUS 18902 x 18911) using RFLP markers (prefixed by -X), isozymes, the *Nor D3* and a slem rust resistance (SrX) loci. Linkage, genetic distance and map ordcr of the loci was computed using the MAPMAKER program (Lander et al. 1987). The broken lines on the map show regions of loose linkage. The map orientation from top to bottom correspond with the short and long arm regions respcctively, except for chromosomes 3D and 4D. The genetic maps on the left side of chromosomes 2D and 3D wcre from different T. *tauschii* progeny. Corresponding positions of other RFLP markers (not detcrmined in AUS 18902 x 18911) are shown with an arrow on the left side of each map.

of resistance to CCN and the root knot nematode *(Meloidogyne javanica)* have been identified (see Table 1), of which the latter is located on chromosome 6D (Kaloshian et al. 1991). However, the gene (or genes) controlling resistance to CCN found in *T.tauschii* is different from the *M. javanica* resistance (Eastwood et al. 1991). A putative locus for the H. *avenae* resistance in one of the accessions of T. *tauschii* is mapped to chromosome 2D (Fig. 2).

Isolation and Application of a D Genome Amplified Sequence

In the course of screening the genomic clones of T. *tauschii,* a D genome amplified sequence (Dgas csIH 44, 2.3kb long) was identified (Lagudah et al 1991a). The "Dgas csIH 44" sequence defines a family of sequences that are highly amplified (300-400 copies) in the D genome and are present on each arm of the chromosomes. The "Dgas" sequence can be detected in some of the Triticeae species, only after long exposure of the autoradiographs using the original 2.3kb *Pst I* genomic clone (Fig. 3). Further analysis of different regions of the "Dgas" clone showed the Hind III fragment - "Dgas 44-3"- (Fig. 3) to be highly specific for the D genome. An application of the Dgas sequence, is in the characterization of introgressed D genome segments in tetraploid wheat as well as triticales containing D genome chromatin.

Several regions of the genetic map of the D genome are deficient for RFLP markers (Fig. 2). These low density marker regions can be mapped using clones generated from other laboratories or by devising strategies for enriching DNA sequences targetted to unsaturated regions of D genome chromosomes. The availability of genetic stocks carrying individual D genome disomic substitutions for their corresponding A and B homoeologues in tetraploid wheat (Joppa 1988; Joppa and Williams 1988) are a valuable genetic resource in enriching for chromosome specific regions. The use of Pulse Field Gel Electrophoresis to analyze large DNA fragments of these genetic stocks and probed with the "Dgas" sequence will enable the detection and

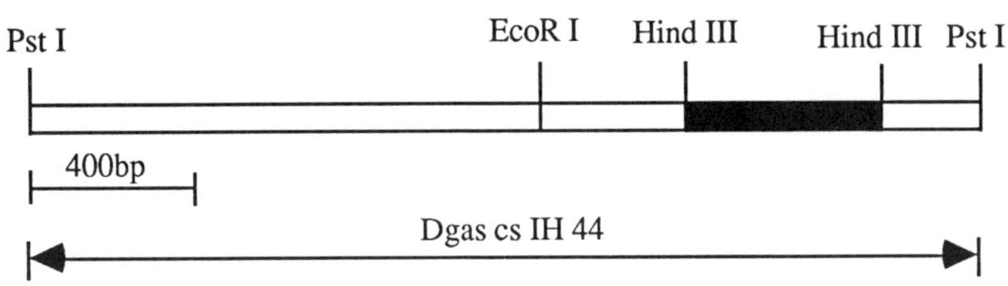

Fig. 3. A partial restriction map ol the D genome amplified sequence (Dgas cs IH44). The Hind III fragment (44-3) is highly specific for the D genome.

isolation of regions enriched for specific D genome chromosomes (as suggested by May and Appels 1987a). Further analysis of the flanking regions of the DNA fragments detected by the "Dgas" sequence and in combination with sequences delineating non-saturated intervals in the genetic maps, can be used to isolate potential clones specific for the targetted interval.

GENETIC DIFFERENTIATION IN THE DIPLOID RELATIVE TO THE HEXAPLOID D GENOME

The degree of differentiation in T. *tauschii* affects the level of homology with the D genome of wheat. Consequently the source of T. *tauschii* used in introgressing a trait into wheat may limit or enhance the removal of undesirable chromatin associated with the target region. There is evidence

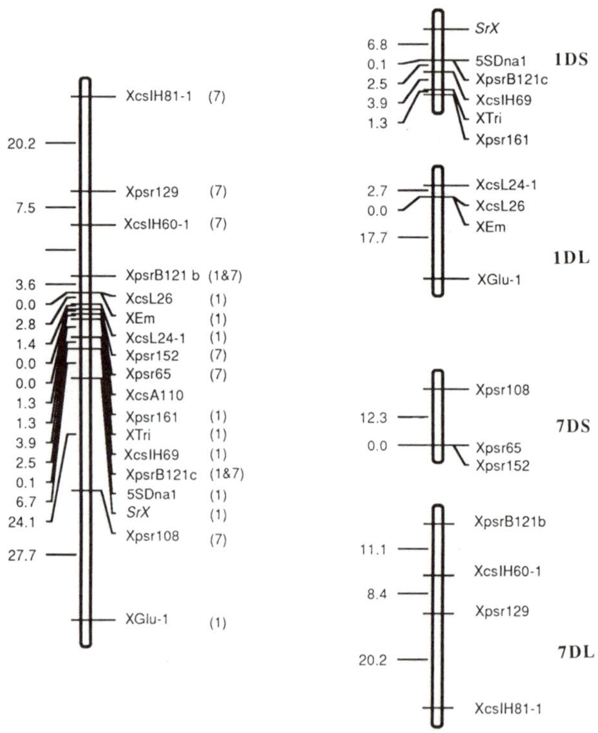

Fig. 4. (A). Composite linkage group of proximal markers on chromosome 1 and 7. (B). The linkage group obtained from truncated data set of markers based on the short (S) and long (L) arm locations. The composite map represent the single apparent 'order' with highest likelihood produced by the MAPMAKER algorithm. Distances in centimorgans are shown on the left side of the maps. The order of loci in the truncated data set *(B) is* similar to those obtained from other *T.tauschii* progeny where linkage of markers on chromosomes 1 and 7 were independent (Lagudau et al. 1991a).

for differential chromosome pairing between chromosome 7D and its equivalent homologue from T. *tauschii* substitution lines into wheat derived from ssp. *eusquarrosa* vars. *typica* and *anathera,* and ssp. *strangulata* var. *strangulata.* The greatest reduction in heterologous pairing occured with var. *typica* and least with var. *strangulata* (Penner et al. 1986). Substitution of chromosome 6D of T. *tauschii* into wheat resulted in reduced pairing with its homologue in *T. aestivum* compared to that obtained with intervarietal hybrids of T. *aestivum* (Dvorak 1988). Differential levels of homologous chromosome pairing observed in intervarietal hybrids of cultivated wheat has been attributed to non-structural differentiation between homologues in wheat varieties (Dvorak and McGuire 1981). Similar non-structural genetic changes in the chromatin of the diploid D genome may be responsible for the variable pairing reported relative to chromosomes 6D and 7D of T. *aestivum.*

In the F2 progeny of a cross between two T. *tauschii* accessions (AUS 18902 var. *typica x* AUS 18911 var. *meyeri),* analysis of cosegregation of RFLP markers revealed an unusual association between several markers on chromosome lD and 7D to produce a composite linkage group (Fig. 4). Linkage analysis of the truncated data sets taken from the composite linkage group based on chromosome arm locations of the markers, showed the order of genetic loci to be the same as that obtained in other T. *tauschii* (Fig. 2) and T. *aestivum* F2 progeny, where no linkages were detected between chromosomes 1 and 7 (Chao et al. 1989; Lagudah et al. 1991a). Structural chromosomal changes commonly manifested as translocations have been reported in several wheat varieties (Schlegel and Schlegel 1989). Translocations detected in the D genome are, lD/6D, 3D/5D, 3D/4D, and 6D/7D. However no report has been made about the occurrence of a lD/7D translocation in *T. aestivum.* Karyotypic analysis of the parental T. *tauschii* lines, AUS18902 and AUS18911, revealed cells of the former possessed a reciprocal translocation involving arms of chromosomes 1D and 7D (Hohmann, Lagudah and Appels, unpublished). Similar observations have been made involving these two chromosomes in other T. *tauschii* accessions (B.S. Gill, personal communication). Thus, certain genotypes of the diploid D genome contain certain structural changes in their chromatin not present in the D genome of bread bread wheat. The availability of cytogenetic stocks of wheat (Sears 1954,1966) that enable prior chromosomal location of genetic markers facilitated the detection of the composite linkage group. However, in several crop plants, where genetic maps are currently being produced, well characterized cytogenetic stocks are not available; and hence, pseudolinkages of the type observed in the composite map of chromosomes lD and 7D of T. *tauschii* are likely to remain undetected.

ACKNOWLEDGEMENTS

The excellent technical support of Sutha Chandramorhan is greatly appreciated. We are also grateful to many others involved in the project, particularly, R. Eastwood, L. Hogie, R. Loughman, R.A. McIntosh, C. May, R. Munns and R. Wilson. Most of the studies reported in this paper was supported by grant # CSP35W from the Grains Research and Development Corporation, Australia.

REFERENCES

Appels, R., and Dvorak, J. 1982. The wheat ribosomal DNA spacer region: its structure and variation in populations and among species. *Theor. Appl. Genet.* 63:337-348.

Appels,R. and Lagudah, E.S. 1990. Manipulation of chromosomal segments from wild wheat for the improvement of bread wheat. *Aust. J. Plant Physiol.* 17:253-266.

Asiedu, R., Fischer, J.M., and Driscoll, C. 1990. Resistance to *Heterodera avenae* in the rye genome of triticale. *Theor. Appl. Genet.* 79: 331-336.

Baker, B., Schell, J., Lorz, H. and Fedoroff, N. 1986. Transposition of the maize controlling element *Activator* in tobacco, *Proc. Natl Acad. Sci USA.* 83: 4844-4848.

Balcells, L., Swinburne, J., and Coupland, G., 1991, Transposons as tools for the isolation of plant genes, *TIBTECH. 9:* 31-37.

Baum, M.B. and Appels, R. 1991. The cytogenetic and molecular architecture of chromosome lR - one of the most widely used sources of alien chromatin in wheat varieties. *Chromosoma* 101,1 - 10.

Chao, S., Sharp, P.J., Worland, A.J., Warham, E.J., Koebner, R.M.D. and Gale, M.D. 1989. RFLP-based genetic maps of wheat homoeologous group 7 chromosomes. *Theoretical and Applied Genetics.* 78: 495-504.

Chevre, A.M., Jahier, J. and Trottlet, M. 1989. Expression of disease resistance genes in amphiploids- Triticum tauschii (Coss) Schmal. *Cereal Res Commun.* 17: 23-29.

Conner, R.L., MacDonald, M.D., Whelan,E.D.P. 1988. Evaluation of take-all resistance in wheat-alien amphiploid and chromosome substitution lines. Genome 30:597-602.

Cox, T.S., Hatchett, J.H., Gill, B.S., Raupp, W.J., and Sears, R.G. 1990, Agronomic performance of hexaploid wheat lines derived from direct crosses between wheat and Aegilops squarrosa, *Plant Breed.* 105: 271-277.

Dundas, I.S., Shepherd, K.W. and Fisher, J.M. 1990. Prospects for breeding wheat resistant to the cereal cyst nematode using a gene from rye chromosome 6R, In Proc. of 6th Assembly, Wheat Breeding Society of Australia, 147-149.

Dvorak, J. 1988. Cytogenetical and molecular inferences about the evolution of wheat. *Proc. 7th Int. Wheat Genet. Symp.*, Cambridge, England. pp. 187-92.

Dvorak, J. and McGuire, P.E. 1981. Nonstructural chromosome differentiation among wheat cultivars with special reference to differentiation of chromosomes in related species. *Genetics, 97,* 391-414.

Eastwood R.F., Lagudah, E.S., Appels,R., Hannah,M., and Kollmorgen J.F. 1991. *Triticum tauschii:* A novel source of resistance to cereal cyst nematode *(Heterodera avenae). Aust J. Agric. Res.* 42, 69-77.

Gale, M.D., Chao, S. & Sharp, P.J. 1990. RFLP mapping in wheat - Progress and problems. In *Gene manipulation in plant improvement. 11,* ed. J.P. Gustafson, pp. 353364. New York: Plenum Press.

Gill, B.S. and Raup, W.J. 1987. Direct genetic transfers from *Aegilops squarrosa* L. to hexaploid wheat. *Crop Sci.* 27, 445-450.

Gill, B.S., Hatchett,J.H., Cox, T.S., Raupp, W.J.,Sears, R.G., Martin, T.J. 1986. Registration of CS85WGRCOI hessian fly-resistance hard red winter wheat germplasm. *Crop Sci.* 26, 1266-1267.

Gill, B.S., Sharma, H.C., Raupp, W.J., Browder, L.E., Hatchett, J.H., Harvey,T.J., Moseman, J.G., and Waines, J.G. 1985. Evaluation of Aegilops species for resistance to wheat powdery mildew, wheat leaf rust, Hessian fly and greenbug. *Plant Dis.* 69, 314-316.

Gill, K.S., Lubbers, E.L., Gill, B.S., Raupp, W.J., and Cox, T.S. 1991. A genetic linkage map of *Triticum tauschii* (DD) and its relationship to the D genome of bread wheat (AABBDD). *Genome,* 34: 362-374.

Gorham,J., Hardy, C., Wyn Jones, R.G., Joppa, L.R., and Law, C.N. 1987. Chromosomal location of a K/Na discrimination character in the D genome of wheat. *Theor. Appl Genet.* 74: 584-588.

Halloran, G.M. 1968. Wheat collecting expedition to Afganistan. *Proc. 3rd Int. Wheat Genet. Symp.* Canberra, Australia, 159-160.

Hart,G.E., 1987. Genetic and biochemical studies of enzymes. In *Wheat and wheat improvement,* ed. E.G. Heyne, pp. 199-214. Madison: American Society of Agronomy.

Harvey,T.L., Martin, T.J., Livers, R.W. 1980. Resistance to biotype C greenbug in synthetic hexaploid wheats derived from *Triticum tauschii, J. Econ. Entomol.* 73, 387-389.

Hatchett, J.H., Gill, B.S. 1981. D-genome sources of resistance in T. *tauschii* to Hessian *fly. J. Hered.* 72, 126-127.

Islam, A.K.M.R., Shepherd, K.W. & Sparrow, D.H.B. 1981. Isolation and characterisation of euplasmic wheat-barley chromosome addition lines. *Heredity* ,46: 161-174.

Joppa, L.R. 1980. Inheritance of resistance to greenbug toxicity in an amphiploid of *Triticum turgidumlT. tauschii. Crop Sci.* 20, 343-345.

Joppa, L.R. 1988. Cytogenetics of tetraploid wheat. In *Proceedings of the Seventh International Wheat Genetics Symposium,* ed. T.E. Miller & R.M.D. Koebner, pp.197202, Cambridge: Institute of Plant Science Research, Cambridge Laboratory.

Joppa, L.R. & Williams, N.D. 1988. Langdon durum disomic substitution lines and aneuploid analysis in tetraploid wheat. *Genome,* 30, 222-228.

Kaloshian, I., Roberts, P.A., Thomason, I.J. 1989. Resistance to *Meliodogyne* spp. in allohexaploid wheat derived from *Triticum turgidum* and *Aegilops squarrosa. J.* Nematology 21, 42-47.

Kaloshian, I., Waines, J.G., Roberts, P.A., and Thomason, I.J. 1991. Chromosomal location of root-knot resistance gene in the D genome of wheat, J. *~ered.* 82: 254-255.

Kerber, E.R. 1983. Suppression of rust resistance in amphiploids of *Triticum. Proc. 6th Int. Wheat Genet Symp.* Kyoto, Japan. pp. 813-817.

Kerber, E.R. 1987. Resistance to leaf rusts in hexaploid wheat: Lr32, a third gene derived from T. tauschii. Crop Sci. 27, 204-206.

Kerber, E.R., Dyck, P.L. 1969. Inheritance in hexaploid wheat of leaf rust resistance and other characters derived from *Aegilops squarosa. Can.* J. *Gen. Cytol.* 11, 639-6471.

Kerber, E.R. and Dyck, P.L. 1978. Resistance to stem and leaf rust of wheat in *Aegilops squarrosa* and transfer of a gene for stem rust resistance to hexaploid wheat. *Proc. 5th Int. Wheat Genet. Symp.* New Delhi, India, 358-364.

Kerber, E.R and Green, G.J. 1980. Suppression of stem rust resistance in hexaploid wheat cv. Canthatch by chromosome 7DL. *Can. J. Bot.* 58: 1347-1350.

Kihara, H. 1944. Discovery of the DD analyser, one of the ancestors of T. *vulgare. Agric. Hort. 19,* 889-890.

Kihara, H., Yamashita, K., and Tanaka, M. 1965. Morphological, Physiological, geographical and cytological studies in Aegilops and Triticum collected in Pakistan, Afghanistan, Iran. Results of the Kyoto University Scientific Expedition in 1955 1:11 18.

Lagudah, E.S. and Halloran, G.M. 1988. Phylogenetic relationships in *Triticum tauschii* the D genome donor to hexaploid wheat. 2. Inheritance and chromosomal mapping of the HMW subunits of glutenin and gliadin gene loci of T. *tauschii. Theor. Appl. Genet.* 75 :599-605.

Lagudah, E.S., and Halloran, G.M. 1989. Phylogenetic relationships of *Triticum tauschii* the D genome donor to hexaploid wheat. 3. Variation in, and the genetics of, seed esterases *(Est-S). Theor. Appl. Genet.* 77: 851-856.

Lagudah, E.S., Appels, R., Brown,A.H.D., and McNeil, D. 1991a. The molecular-genetic analysis of *Triticum tauschii* - the D genome donor to hexaploid wheat. Genome 34, 375 - 386.

Lagudah, E.S., Appels,R. and McNeil, D. I991b. The *Nor-D3* locus of *Triticum tauschii:* natural variation and linkage to chromosome 5 markers. Genome 34, 387 - 395.

Lagudah, E.S., MacRitchie, F., and Halloran, G. M. 1987. The influence of high-molecularweight subunits of glutenin from *Triticum tauschii* on flour quality of synthetic hexaploid wheat. *J. Cereal Sci.* 5: 129-138.

Lander, E.S., and Botstein, D. 1989. Mapping Mendelian factors underlying quantitative traits using RFLP linkage maps. *Genetics* 121: 185-199.

Lander, E.S., Green, P., Abrahamson, J., Barlow, A. Daly, M.J., Lincoln S.E., and Newburg, L. 1987. MAPMAKER: An interactive computer package for constructing primary genetic linkage maps of experimental and natural populations. *Genomics, 1:* 174- 181.

Lubbers, E. L., Gill, K.S., Cox, T.S., and Gill, B.S. 1991. Variation of molecular markers among geographically diverse accessions of *Triticum tauschii, Genome,* 34: 354-361.

May, C.E., and Appels, R. 1987a. The molecular genetics of wheat: toward an understanding of 16 billion base pairs of DNA. In *Wheat and wheat improvement* ed. E.G. Heyne, ppl65-198 . Madison: American Society of Agronomy.

May, C.E., and Appels, R. 1987b. Variability and genetics of spacer DNA sequences between the ribosomal RNA genes of hexaploid wheat *(Triticum aestivum). Theor. Appl. Genet.* 74: 617-624.

May, C.E., and Appels, R. 1988. Allelism of the nucleolus organiser regions of hexaploid wheat. *Proc. 7th Int Wheat Genet Symp.* Cambridge, UK, pp. 577-583.

May, C.E. and Lagudah, E.S. 1992. Inheritance in hexaploid wheat of *Septoria tritici* blotch resistance and other characters derived from *Triticum tauschii, Aust. J. Agric. Res.* 43: in press

McFadden, E.S. and Sears, E.R. 1944. The artificial synthesis of *Triticum spelta. Rec. Gen. Soc. Am.* 13: 26-27.

McFadden, E.S. and Sears, E.R. 1946. The origin of *Triticum spelta* and its free-threshing hexaploid relatives. *J. Hered* . 37, 81-89; 107-116.

Merkle, O.G. and Starks, K.J. 1985. Resistance of wheat to yellow sugarcane aphid *(Homoptera: Aphididae). J. Econ. Entomol.* 78:127-128.

Multani, D.S., Dhaliwal, H.S., Singh, P., Gill, K.S. 1988. Synthetic amphiploids of wheat as a source of resistance to Karnal bund (Neovossia indica). *Plant Breeding* 101:122125.

Nakai, Y. 1979. Isozyme variations in *Aegilops* and *Triticum IV.* The origin of the common wheats revealed from the study on esterase isozymes in synthesized hexaploid wheats. *Jpn. J. Genet.* 54, 175-189.

Nakai, Y. 1981. D genome donors for *Aegilops cylindrica* (CCDD) and *Triticum aestivum* (AABBDD) deduced from esterase isozyme analysis. *Theor. Appl. Genet.* 60, 11-16.

Nishikawa, N. 1983. Species relationship of wheat and its putative ancestors as viewed from isozyme variation. Proc. 6th Int. Wheat Genet. Symp., Kyoto, Japan: 59-63.

Nkongolo, K.K., Quick, J.S., Limin, A.E., and Fowler, D.B. 1991. Sources and inheritance of resistance to Russian wheat aphid in *Triticum* species amphiploids and *Triticum tauschii, Can. J. Plant Sci.* 71:703-708.

Payne, P.I. 1987. Genetics of wheat storage proteins and the effect of allelic variation on bread-making quality. *Annual Review of Plant Physiology,* 38, 141-153.

Payne, P.I., Holt, L.M., Worland, A.J., and Law, C.N. 1982. Structural and genetic studies on the high- molecular-weight subunits of wheat glutenin. 3. Telocentric mapping of the subunit genes on the long arms of the homoeologous group lchromosomes. *Theor. Appl. Genet.* 63: 129-138.

Penner, G.A., Kerber, E.R., and Larter, E.N. 1986. The differentiation between chromosome 7D of *Triticum aestivum cv.* Canthatch and *Triticum tauschii* as measured by chromosome pairing. *Can. J. Genet. Cytol.* 28: 385-389.

Rivoal, R., Dosba, F., Jahier, J., and Pierre, J.S. 1986. Les lignees d'addition *ble-Aegilops ventricosa* Tausch: VI. Etude de la localisation chromosomique de la resistance a l'egard *d'Heterodera avenae* Woll. *Agronomie,* 6: 143-148.

Roberts, P.A., Van Gundy, S.D. and Waines, J.G. 1982. Reaction of wild and domesticated *Triticum* and *Aegilops* species to root-knot nematodes (Meloidogyne). *Nematologica* 28: 182-191.

Rowland, G.G. and Kerber, E.R. 1974. Telocentric mapping in hexaploid wheat of genes for leaf rust resistance and other characters derived from *Aegilops squarrosa. Can.J. Genet. Cytol.* 16: 137-144.

Schlegel, G. and Schlegel, R. 1989. A compendium of reciprocal intervarietal translocations in hexaploid wheat. *Kulturpflanze.* 37: 163-176. Sears, E.R.

1954. The aneuploids of common wheat. *Mo. Agric. Exp. Stn. Res. Bull.* 572, 1-59.

Sears, E.R. 1966. Nullisomic-tetrasomic combinations in hexaploid wheat. In: *Chromosome Manipulations and Plant Genetics.* Ed. R. Riley & K.R. Lewis,pp.29-45. Edinburgh: Oliver and Boyd.

Sears, E.R. 1981. Transfer of alien genetic material to wheat. In *Wheat Science-today and tomorrow,* ed. L.T. Evans & W.J. Peacock, pp.75-89. Cambridge: Cambridge University Press.

Schachtman, D.P., Munns, R., and Whitecross, M.l., 1991, Variation in sodium exclusion and tolerance in *Triticum tauschii, Crop Sci.* 31: 992-997

Shebeski, L.H. 1958. Speculation on the impact of the D genome. *Proc. Ist Int. Wheat Genet. Symp.* Winnipeg, Canada. pp. 237-241.

Singh, N.K., Shepherd, K.W. 1988. Linkage mapping of genes controlling endosperm storage proteins in wheat 2. Genes of the long arms of group I chromosomes. *Theor. Appl. Genet.* 75:642-650.

Tanksley, S.D., Young, N.D., Paterson, A.H. and Bonierbale, M.W. 1989. RFLP mapping in plant breeding: new tools for an old science. *Bioltechnology* 7:257-264.

Thomas, J.B. and Conner, R.L. 1986. Resistance to Colonization by the wheat curl mite in *Aegilops squarrosa* and its inheritance after transfer to common wheat. *Crop Sci. 26,* 527-530.

UNDERSTANDING AND EXPLOITING THE TRANSCRIPTIONAL CONTROL MECHANISMS OF PLANT CELLS

Michael Bevan, Michael Holdsworth, Diane Shufflebottom,
Claire Nall, Marta de Torres Zabala, Keith Edwards[1]
and Wolfgang Schuch[1].

Cambridge Laboratory, IPSR, Colney Lane, Norwich, UK.
[1]ICI Seeds plc, Jealotts Hill, Bracknell, Berks, UK.

INTRODUCTION

The generation and maintenance of plant form is brought about by the sequential activation or repression of the transcription of specific genes in time and space. The mechanisms governing this particularly complex process are thought to be common to many eukaryotes such as yeast, plants and vertebrates, and the extensive research aimed at understanding how patterns of gene expression are initiated and maintained is of compelling interest and importance. Evidence from many different types of experiments in diverse organisms showed that regions of genes involved in regulating transcriptional initiation, the primary control point for RNA polII activity are usually located 5' of the coding region. This region can be divided further into two distinct regions: one that determines the exact start point for transcriptional activation and the basal level of transcription, and another distal region that confers the specificity and the magnitude of transcription (reviewed by Lewin, 1991). Both regions act by assembling specific combinations of proteins called transcription factors, some of which bind DNA and other proteins, and some of which are involved only in protein-protein interactions. transcription factors can be classified according to their structure, eg. leucine zipper, zinc finger, homeodomain etc, and most characterised factors appear to be composed of several domains that are interchangeable (reviewed by Frankel and Kim, 1991). There is a DNA binding motif, a region necessary for protein-protein interactions, for example a leucine zipper, and an activation domain that is thought to interact with the basal transcription apparatus to stimulate transcription. Together, complexes of transcription factors mediate the activity of RNA polII on that gene via their interactions with the basal transcription machinery. The final pattern and degree of expression of a gene is thought to be due to the combined effects of several different transcription factors that interact specifically with DNA sequences and with each other to bring about a novel pattern of expression. An increased understanding of the mechanisms of transcriptional activation will have several important practical consequences for the improvement of plants for agriculture. First, the transcriptional machinery of a gene is the penultimate step in signal transduction systems that transduce all manner of signals, including responses to the environment, therefore a clear idea of the mechanism of a signal transduction pathway needs knowledge of the last steps of transcriptional activation. A plant's responses to pathogens, environmental cues such as daylength, and the efficiency of storage product accumulation are a few of the important processes involving signal transduction pathways that need to be understood before desired alterations could be made. Secondly, once the "combinatorial code" has been deciphered (that is, which combinations of transcription factors give rise to particular patterns of expression), it will be possible to construct transcription units with predictable and desired patterns of expression. Presently much effort is expended on finding endogenous promoters to suit particular expression problems. Thirdly, transcriptional control is a common mechanism for coordinately regulating sets of genes, for example, those encoding enzymes of pathways. Understanding and manipulating a common transcriptional control elements will cause, in theory, altered expression patterns of sets of genes in similar ways. This would be a great advantage in altering the flux through pathways or coordinating the expression of sets of novel genes. For example, it has demonstrated recently that an alteration to the oncoprotein c-Jun, involving a deletion of its activation domain, suppressed oncogenesis by acting as a dominant suppressor of normal c-JUN function (Lloyd et al., 1991). Theoretically similar changes could be made to plant transcription factors that could suppress activity of entire sets of genes. Finally, one could envisage designing new signal transduction pathways that would make plants respond to entirely new extrinsic signals, such as chemicals, or in response to altered environmental conditions.

All of these possible ways of manipulating plant metabolism and development are fairly obvious, but the information needed to perform these tasks has only just started to be gathered, and detailed knowledge of transcriptional control of several gene systems will need to be gathered before general rules of value in predicting promoter behaviour from DNA sequence. While many of the principles involved will be established from work in animals and yeast, important aspects unique to plants will almost certainly be uncovered in this type of work. In this report we describe progress in our work on understanding two gene expression sytems, both of which are important in agricultural production.

1. MECHANISMS UNDERLYING THE DIVERSITY OF TRANSCRIPTIONAL REGULATION OF PAL GENES

Introduction

Phenylpropanoids are used in the elaboration of many different plant secondary products, such as anthocyanin pigments, some of the anti-pathogen phytoallexins, and lignin and other polyphenolics (Hahlbrock and Scheel, 1989). These compounds are needed in different amounts in different cell types at different times, therefore the biosynthetic pathway must be responsive to these complex demands. There is substantial evidence that levels of the enzymes in the pathway are modulated to a significant degree by transcriptional regulation of the congnate genes, indicating that the transcription of the genes is complex and tightly controlled (Lamb et al., 1989). The first enzyme in the pathway is phenylalanine ammonia-lyase, which converts phenylalanine to cinammic acid, the precursor to all phenylpropanoids. The enzyme(s) are encoded by small multigene families composed of 3-4 members in *Phaseolus*, parsley (*Petroselinium crispum*) and *Arabidopsis*. It is possible that the diverse patterns of expression are derived from different genes with individual patterns of expression, or that an individual gene has many possible modes of expression. The latter possiblity has been demonstrated in the case of the *Phaseolus* gPAL2 gene by Bevan et al. (1989) and Liang et al. (1989) which was shown to be expressed in response to light, mechanical damage and pathogen ingress, in developing xylem cells, pigmented tissue and in meristematic tissues. A similar complex pattern of expression has been observed for both parsley 4-coumarate CoA ligase genes by Hauffe et al. (1991) and Douglas et al. (1991) and for the genes encoding chalcone synthase (CHS) (Schmid et al. 1990). It is likely that the basis for both the differences in expression patterns between members of the gene family and for the complex patterns of expression demonstrated by individual genes involves multiple *cis*-acting DNA sequences in the promoter and their interaction with multiple cognate transcription factors (eg, as exemplified by the murine a fetoprotein gene promoter, Hammer et al., 1987). The aim of much of the present effort in this field is to delineate these *cis* sequences and use these to obtain the cognate transcription factors. Once these have been obtained, suitable methods, such as *in vitro* transcription systems, can be used to study interactions between factors and between factors and the basal transcription machinery. In this review we present an interim report of our progress towards these goals using the gPAL2 promoter of *Phaseolus*.

Results

Our previous work had demonstrated a complex pattern of expression from a gPAL2 promoter of about 1000 bp, fused to a GUS reporter gene and transformed into tobacco and potato plants (Bevan et al., 1989). High levels of expression were observed in developing xylem tissue, surrounding the sites of mechanical damage, in floral tissues including pigmented regions of the petal, in meristematic tissue such as the root tip and apical meristem, and in epidermal cells in response to light. In an attempt to separate the putative *cis* sequences conferring these patterns of expression, a series of unilateral and internal deletions of the promoter were made, fused to the GUS reporter gene and introduced into tobacco and potato. PAL sequences encoding 18 amino-terminal amino acids of PAL were fused in-frame to a GUS reporter gene in the transforming vector pBI 101.2 (Bevan, 1984). The structure of these is shown in Figure 1.

About 10-20 independent transformants were assayed for each construct using histochemical methods (Jefferson et al., 1987). The minimum fully active promoter (Δ15) extends to -253 bp from the transcription start site, and smaller promoters comprising DNA extending to -123 and -75bp do not express GUS at detectable levels. It must be noted that these assays were histochemical and can therefore only be semi-quantitative. It is possible that the deletion of an AT tract just distal to the Δ15 endpoint may lower expression, as these sequences are known to bind HMG-like proteins that generally enhance transcription, possibly by altering chromatin structure (Johns, 1987). We detected no specific loss of function using unilateral deletions, but rather a sudden loss of expression in all tissues in deletions proximal to -253bp. Unilateral deletions within the -1000bp promoter reveal a more complex promoter structure however.

A small deletion (Δ5) extending from -123 to -151, had no effect on expression in all tissue types examined (Table 1), but more extensive deletions 5' to -194 and -245 successively weakened and nearly abolished expression in xylem tissue and pigmented regions of tobacco petals and in the trichome hairs, while still allowing expression in pollen, root tips and cortical cells. Xylem, trichome and petal expression was recovered in deletions spanning more distal regions such as Δ8 - Δ13, indicating that sequences between approximately -151 and -253bp were necessary for expression in these tissues. Tuber wound periderm expression could be abolished by more extensive proximal deletions from -75bp (Δ2, Δ3), but was still evident in deletions that abolished expression in xylem, petals and trichomes (eg. Δ6). Expression in mature pollen and in cortical cells was maintained in all internal deletions tested, indicating that elements responsible for directing expression in these tissues resided in several locations in the regions deleted or in other regions not tested. Root tip expression was evident in all internal deletions, indicating a possible redundancy of elements dispersed throughout the promoter regions tested or present in other regions not tested. However, root tip expression was abolished by reversing the orientation of a -380 to -63bp fragment fused to a minimal -40bp CaMV 35S promoter, indicating a reliance on the absolute orientation of element(s) betwen these coordinates for activity. The elements necessary for this could be further confined to a region between -63 and -110bp, as PCR1 and PCR2 fusions to the CaMV promoter did not express in the root tip. It is possible that these elements could also be between -212 and

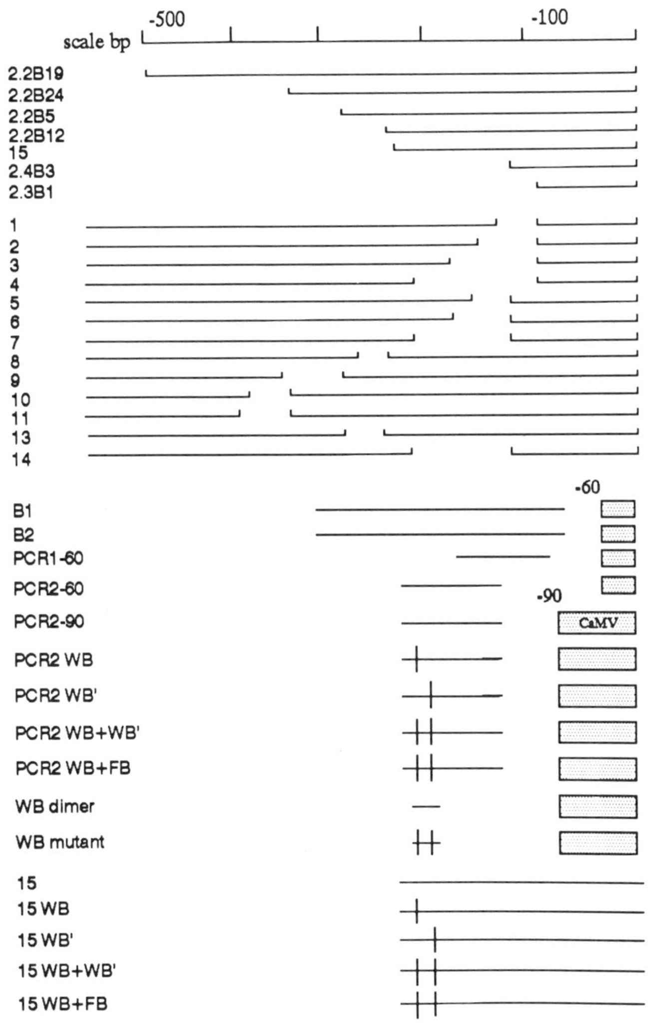

Fig. 1. The structure of PAL promoter-GUS fusions used. Regions of the promoter are drawn to scale and are represented by lines. Vertical lines represent the positions of mutations described in Figure 3.

Table 1. The activity of many of the deletions of the PAL promoter described in Figure 1 was measured histochemically in the cell types described in multiple transformants of tobacco and potato. The score is based on observations of several samples from 10-20 independent transformants.

PROMOTER		TISSUE TESTED						
Name	Co-ordinates	Xylem	Cortex	Root	Petals	Hair Cells	Wounded Tuber Periderm	Pollen
2.2B24	-420	+	+	+	+	+	+	+
2.2B5	-342	+	+	+	+	-	+	+
2.2B12	-283	+	+	+	+	-	+	+
Δ15	-253	+	+	+	+	-	+	+
2.4B3	-123	-	-	-	-	-	-	-
2.3B1	-75	-	-	-	-	-	-	-
1	-75 to -115	+	+	+	+	+	+	+
2	-75 to -151	+	+	+	+	+	VW	+
3	-75 to -194	W	+	+	W	-	VW	+
4	-75 to -245	VW	+	+	VW	-	VW	+
5	-123 to -151	+	+	+	+	+	+	+
6	-123 to -194	W	+	+	W	-	VW	+
7	-123 to -245	VW	+	+	VW	-	+	+
8	-283 to -311	+	+	+	+	+	+	+
9	-342 to -430	+	+	+	+	+	NT	+
10	-420 to -455	+	+	+	+	+	NT	+
11	-420 to -480	+	+	+	+	+	NT	+
13	-283 to -336	+	+	+	+	+	+	+
14	-123 to -250	VW	+	+	VW	-	W	+
B1	-47 to -380	+	+	-	+	+	NT	+
B2	-380 to -47	+	+	+	+	+	NT	+
C13	-380 to -63	+	+	+	+	+	NT	+
C15	-63 to -380	+	+	-	+	+	NT	+
PCR2	-252 to -150	W	-	-	+	NT	NT	+
PCR1	-212 to -110	VW	-	-	+	NT	NT	+
-40CaMV	-	-	-	-	-	-	-	-

KEY: NT = NOT TESTED; W = WEAK EXPRESSION; VW = VERY WEAK EXPRESSION

-380bp, but a deletion in this region (Δ8) did not abolish expression in root tips. Possible interactions between different sets of elements conferring root-tip expression in these two areas, or a functional redundancy further cloud interpretation of root tip expression data. Further constructs have been made to test these ideas. To summarize, data from internal deletion experiments defined at least three separable sets of *cis* sequences: those between -151 and -253 are involved in expression in xylem, trichomes, and in pigmented regions of petals; those between -63 and -110 are involved in root tip expression, although there are other sequences involved. Wound-induced expression in tuber periderm tissue is conferred by sequences between -75 and -151, and by sequences between -123 and -194. It appears that there is a substantial redundancy of information in the promoters tested conferring pollen-specific expression, such that all internal deletions demonstrated pollen-specific activity, while this expression was abolished by unilateral deletions to -123. The same interpretation is applied to cortical cell expression (see later). It is important to verify this type of data by reconstructing artificial promoters containing various promoter fragments. This was done in the case of sequences thought to be responsible for xylem specificity. Fusions of PCR1 (-212 to -110) and PCR2 (-252 to -150) with a minimal -60 CaMV promoter directed respectively, very low and low levels of expression in developing xylem cells, and both expressed low level of expression in pigmented regions of tobacco petals. Expression was also seen in pollen, but not in root tips or cortical cells. These results confirmed our findings of important *cis* sequences between -245 and -151 conferring expression in these tissues. It is possible that there are two elements separated at nucleotide -212, conferring xylem specificity, as PCR1 gave weak expression while PCR2 gave consistently higher expression in xylem. It is possible that the element(s) conferring petal-specific expression may be found in the common regions between PCR1 and PCR2, namely nucleotides -252 to -212. The data from this analysis is summarised in Table 1.

We are particularly interested in xylem-specific gene expression and wish to define the precise sequences responsible for conferring this expression. To this end electrophoretic mobility shift assays (EMSA) were performed using radiolabelled PCR2 fragment, as this was able to confer xylem-specific expression on a basal promoter. Nuclear proteins were isolated from stems of 2-3 month old tobacco plants, which were undergoing extensive secondary thickening of the xylem. The majority of the tissue extracted consisted of xylem, with epidermal and phloem tissue comprising the rest. High levels of PAL activity and mRNA are found specifically in the xylem tissues of these mature stems. Preliminary experiments showed that binding activity was present in a 40% ammonium sulphate fraction, and this was further purified by chromatography over heparin-agarose using a linear gradient of KCl to elute bound proteins. Specific binding activities eluted at approximately 80mM and 300-400mM KCl in several preparations, and these were pooled and concentrated. Optimum binding conditions were established at 20mM KCl, 15mM HEPES-KOH pH 7.5, 1mM EDTA. Binding was abolished by low concentrations (≤2mM) of divalent cations such as Mg^{2+}.To date no difference has been observed in the binding activities of these two fractions, and our present idea is that these two fractions are different forms of the same activity. In order to establish the exact site(s) of interaction of the binding activity, methylation interference experiments were carried out. These identified a single binding site on the bottom strand that involved G residues at -221, -218 and -216 (Figure 2). It was observed that the three G residues involved in binding formed part of a repeat structure (Figure 3), and this was repeated as a near-perfect copy proximally in PCR2 at -208bp, although these methylation interference experiments did not detect this binding. It is relevant that this possible second binding site is present in PCR1, and that it is embedded in a near perfect G-box concensus (Guiliano et al., 1988). To verify these observations, the G residues above in both copies of the repeat were changed to T residues by *in vitro* mutagenesis, and the binding of this fragment to partially-purified nuclear proteins was examined using EMSA. A further series of mutations that altered the G box consensus (called FB), were also carrried out to examine possible binding at this site. The results are shown in Figure 4. Panel A shows the binding of partially purified nuclear extract to different probes. Those probes with both copies of the putative binding site altered by G to T transitions bound protein much less avidly than fragments with only one or no sites altered. A dimer of the binding site, and a dimer of the mutant binding site, showed avid binding and no binding respectively. We concluded that binding to both sites occurred, and could occur in the absence of binding to the other site. It is noteworthy that PCR1, which contains a single proximal copy of the binding site, also avidly binds the activity. To confirm these results in a more rigourous test, PCR2 was labelled and excess unlabelled mutated fragments were used as competitors. The results of this test are shown in Figure 4, panel B. Internal controls of self-competition showed that a 25-fold molar excess of PCR was sufficient to compete off all bound forms. The same was found when the binding site dimer (called WB) and PCR1 were used as competitors. However, PCR2 with both binding sites mutated and the dimer of mutated binding site were not as efficient competitors. Cold fragments with a single binding site altered were still efficient competitors. These two sets of results confirmed each other and allowed us to draw the conclusion that there are two binding sites on PCR2 that bind a nuclear protein isolated from material rich in xylem tissue. These sites may be involved in directing xylem-specific gene expression, as PCR1 with one copy of the binding site expressed very low levels of GUS in xylem tissue when fused to the -60 CaMV promoter (Table 1), while PCR2 with two binding sites expressed higher levels of GUS in similar constructs. A direct test of the activity of each of the mutated fragments in planta, and of entire promoters containing the same mutations, is presently underway. The structure of these is shown in Figure 1. With this evidence, the final identity of the xylem-specific elements will be possible. It is curious that similar sequences to WB are not immediately apparent in other promoters which direct expression in xylem tissue, such as 4CL (Hauffe et al. 1991; Douglas et al., 1991), and in an *Arabidopsis* PAL gene (Ohl et al., 1991). However, no functional analysis of these promoters has been carried out. We also intend to test these different sequences for petal-specific expression. It would be particularly interesting to see if the G box-like sequence at -208 bp is involved in conferring

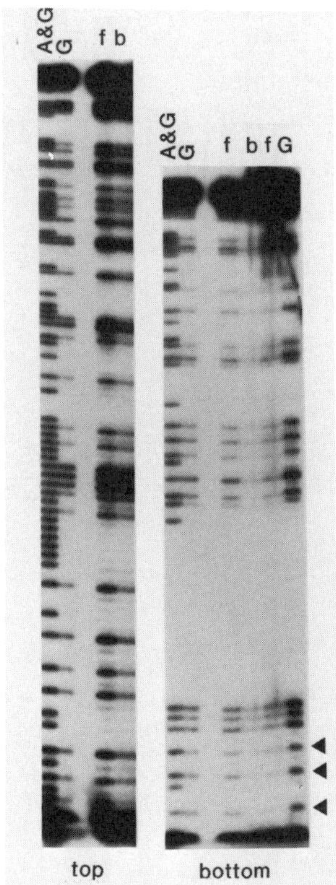

Fig. 2. Methylation interference assay of binding of nuclear proteins from stem tissue to the PCR2 fragment. Top and Bottom refer to the two strands of DNA. A&G and G lanes refer to Maxam-Gilbert reactions of PCR2 that provide markers. The arrows show the three G residues specifically involved in protein binding.

this expression. Similar sequences are found in the chalcone synthase promoter as the box II motif, and these confer elicitor and UV light inducibility in transient assays in parsley protoplasts (Weisshaar et al., 1991). Variants on the sequence 5'-CTCACCTACCA-3' (called the box I motif) are conserved between different PAL promoters and 4CL, and they has been shown to bind proteins in *in vivo* footprinting assays in response to light and elicitor in parsley (Lois et al., 1989). There are two sets of these sequences in GPAL2, one found within the region deleted in Δ1, and the other found in the region deleted by Δ5. These deletions have no effect on xylem expression, but deletion of both elements, such as in Δ2 and Δ4, severly weakens wound periderm-specific expression. No *in vivo* assay for the activity of these box I elements in parsley has yet been carried out, but transient assays in parsley protoplasts have shown that together box I and box II motifs confer light and elicitor-inducible expression.

Discussion

Our aim was to demonstrate the existence of specific *cis* sequences in the PAL promoter, and to begin to understand their functions and interactions that give rise to the complex pattern of developmentally- and environmentally-induced transcription. While not having gained direct evidence as yet implicating the WB and WB' sequences in xylem-specific expression, the present evidence is strongly suggestive of a role. Indeed the principle of multiple *cis* acting sequences has been demonstrated, with

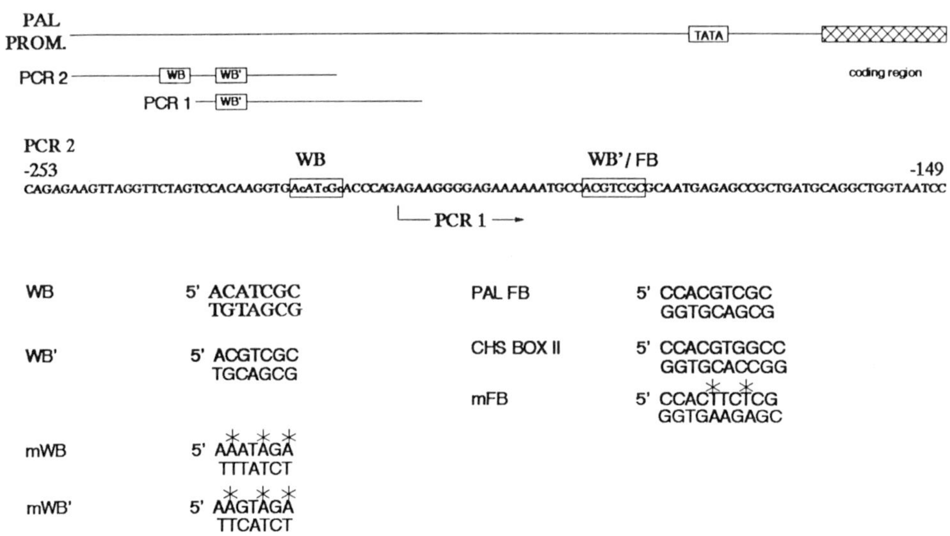

Fig. 3. The sequence of PCR2 is shown, and also its relationship with PCR1 and the entire PAL promoter. The position of the binding site shown in Figure 2 is noted as WB, and the related sequences proximal to this are called WB'/FB. The sequences of these and mutant variants used in EMSA experiments are shown.

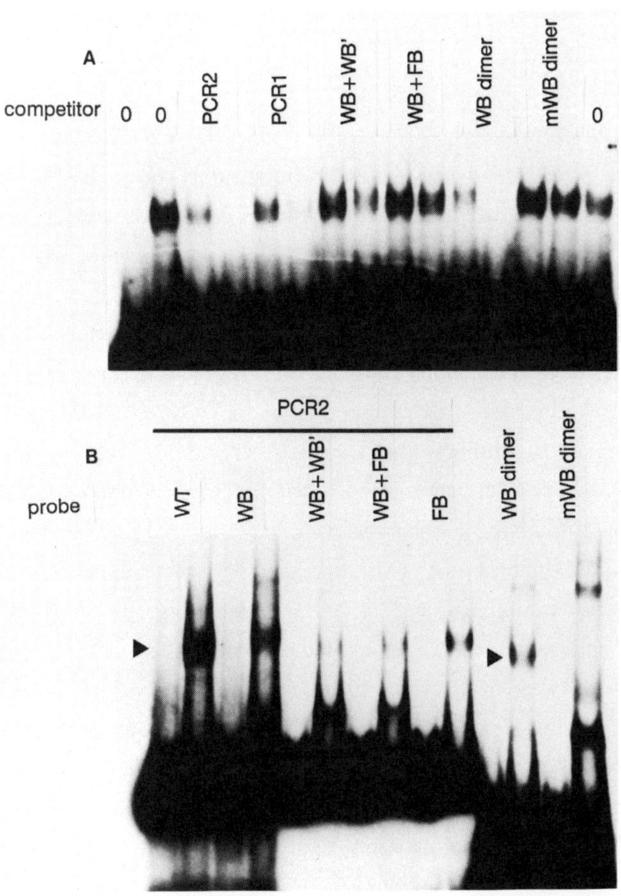

Fig. 4. EMSA of stem nuclear proteins and PCR2 derivatives of the PAL promoter. Panel A shows competition experiments between labelled PCR2 and unlabelled derivatives described in Figure 3. The unlabelled competitors were added at 10- and 100-fold molar excess (from left to right). WB and mWB dimer are dimers of the core binding domain shown in Figure 3. Panel B shows the binding of stem nuclear proteins to labelled probes. WT refers to PCR2, while WB, WB+WB' etc refers to the mutant versions of PCR2. The arrow indicates the specifically retarded band.

separate regions of the promoter conferring xylem and petal-specific expression, root tip expression, wound periderm expression, and pollen and cortical cell expression. Several animal promoters demonstrate this behaviour, such as that of the a fetoprotein gene (Hammer et al., 1987), as does the yeast HIS3 gene (Arndt et al., 1987). Also the CaMV 35S promoter, which demonstrates "constitutive expression" (expression in many cell types at similar relatively high levels), (Benfey et al.,1990), is composed of several distinct *cis* sequences which individually confer expression in the root tip, petals etc and other cell types.The additive or "combinatorial" effect of the activity of these *cis* sequences is to give a constitutive pattern of expression. In the case of the PAL promoter, the combinatorial effects have not yet been assessed, but they give rise to a complicated pattern of expression which may be more than the sum of the individual patterns of gene expression.

2. REGULATION OF A GENE ENCODING A POTATO TUBER PROTEIN.

Introduction

The potato (*Solanum tuberosum* L.) tuber is a major food source in many countries of the world, and subsequently potato has been the target of a good deal of effort directed at engineering disease and herbicide tolerance, and improvements in various crop characteristics. Consequently investigations into the regulation of gene expression in tubers is relevant to these endeavours, as tubers are the main target organ for modification of gene expression. We have been interested in the regulation of genes in tubers for these reasons.

Morphologically tubers are modified stems, which have enlarged radially by limited cell division and substantial expansion. At the molecular level, tuber development is characterised by a massive increase in starch deposition and the synthesis of a limited number of abundant proteins. These include proteinase inhibitors and a 40kd group of proteins called patatin, which are acyl hydrolases. Together these proteins account for over 50% of tuber proteins (reviewed by Bevan, 1991). The synthesis of these proteins has parallels to the synthesis of other somatic storage proteins, especially the VSP proteins of soybean. In both potato and soybean, removal of the sink for these proteins (tubers and pods, respectively) causes deposition in other tissues (Staswick, 1990). It is hypothesised that transcriptional control of the genes encoding these proteins is regulated in part by source-sink relationships of metabolites or other factors. In the case of VSPs, both amino acid levels and jasmonic acid play a major regulatory role (Staswick et al., 1991), while in the case of patatin sucrose or a metabolite of sucrose is involved in this regulation. Promoters from class 1 patatin genes, which are most highly expressed in tubers, can be activated to high levels in many different cell types in potato by exogenous sucrose, (Rocha-Sosa et al.,1989) and to a much lesser extent, even in tobacco (*Nicotiana tabacum*) (Wenzler et al., 1989), which has no homologues of patatin and no record of forming tubers. An attractive hypothesis is that levels of sucrose or some other critical metabolite that senses flux of photosynthate levels can modulate transcription of "tuber-specific" genes. In this way gene expression in sink tissues can be entrained to the flux of metabolites provided by source tissues, in this case source leaves, providing a balance of source and sink strengths. This hypothesis is consistent with the recent finding that the gene(s) encoding the S subunit of ADPglucose pyrophosphorylase (S-ADPGPP) are also up-regulated in tuber tissue, and can be transcriptionally activated in other tissues in response to exogenous sucrose (L. Willmitzer, *pers. comm.*). ADPGPP is a key regulatory enzyme in the flux of photosynthate into starch. Also a gene encoding granule-bound starch synthase is strongly up-regulated by sucrose levels in potato (R. Visser, *pers. comm.*). The tuberous storage root of sweet potato (*Ipomoea batatas* L.) accumulates a major storage protein called sporamin (homologous to Kunitz trypsin inhibitors), and it has recently been shown that the promoter of one of these genes is under the transcriptional control of sucrose in transgenic tobacco plants (Ohta et al., 1991). Thus it may be possible that sucrose or a metabolite of it also coordinates the expression of various other genes that are expressed during tuberisation. In this regard it is relevant to note that proteinase inhibitor genes are sucrose-inducible in leaf tissue (Johnson and Ryan, 1990), as well as being induced during normal tuber development and subsequent to wounding of potato tissues. Finally, the recent finding that several genes of the Calvin Cycle are transcriptionally repressed by sucrose (Sheen, 1990) adds further credence to the arguement that sucrose or one of its metabolites coordinates gene activity in source tissue (the photosynthesising leaf) and sink tissue (the tuber). It is our aim to understand this regulation at a molecular level, to attempt to examine the cellular signals involved, and to see how pervasive and important the control of enzyme levels by metabolites is in plants.

Results

The clearest way to begin the analysis of the effects of sucrose on patatin gene expression is to define the appropriate *cis* sequences involved on the promoter. A series of unilateral deletions of a class 1 promoter have been made by Jefferson et al. (1990) and Liu et al. (1990). Both studies found that the minimal promoter able to confer tuber-specific and sucrose-inducible expression extended to approximately -300bp from the cap site. Also, longer promoters up to -2500bp form the cap site conferred progressively higher levels of expression, indicating that multiple copies of sequences may confer expression in tubers and sucrose inducibility. A possible candidate was a repeat common to all class 1 genes examined (Mignery et al., 1988) that contained an AT-rich motif. Further effects mediated by upstream sequences included a "silencer" region distal to -674bp (Jefferson et al., 1990). In these studies it was not possible to distinguish the effects of exogenous sucrose from tissue-specific patterns of expression (as measured by assaying GUS levels in different organ types of mature glasshouse-grown plants). This is not suprising due to the large fragments used, each of which could contain multiple distinct elements conferring different responses to the promoter. Subsequent experiments started on the assumption that the repeat element situated -224bp distal of the cap site, and repeated at -579bp and -1100bp with minor variation, may be important in conferring sucrose-inducibility and tuber-specificity.

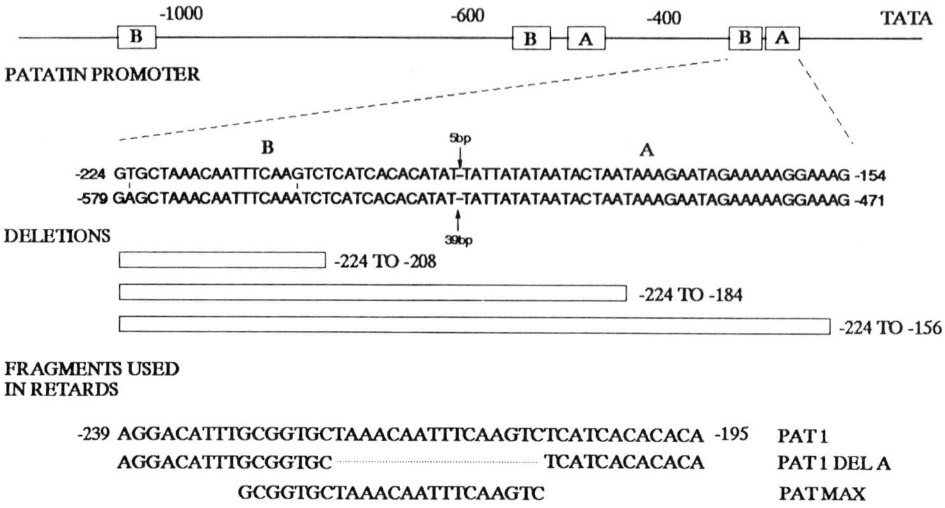

Fig. 5. The structure of the class 1 patatin promoter used in this study is shown. At the top the arrangement of A and B domains of the repeats is shown, with the sequence of the proximal A+B repeat displayed underneath. Parts of the proximal repeat deleted are shown as boxed regions, while under these the sequence of fragments used in EMSA experiments described in Figure 8 is shown.

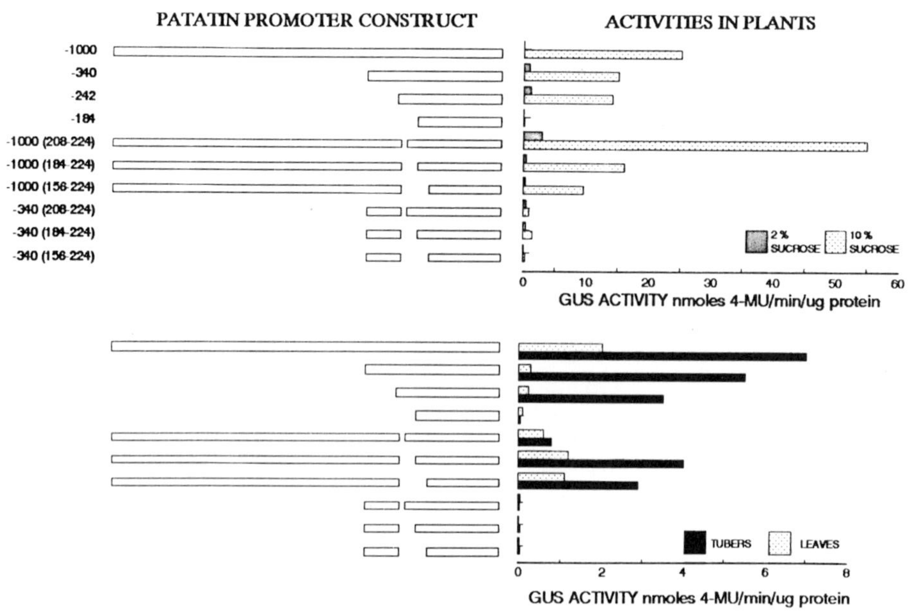

Fig. 6. GUS activities expressed from the different internal deletions of the patatin promoter in transgenic potato. The promoters are shown on the left (not to scale) with the deleted regions missing. The corresponding activities of each promoter type in response to exogenous sucrose (top panel) and leaves and tubers (bottom panel) is shown.

The sequence of these repeats is shown in Figure 5. Two types of experiments were carried out to test this hypothesis. One involved making specific deletions in the repeat in a minimal -360bp promoter and in a -920bp promoter, while in the other, variants of the repeat were fused to truncated 35S promoter derivatives. In all cases the promoters were fused to GUS and transgenic potato plants obtained by *Agrobacterium*-mediated transformation. Sufficient replicates and independent plants were assayed to obtain statistically significant results. The data showed that deletion of a small region of the repeat, from -224 to -208 (Figure 5) which left the AT rich motif intact, almost completely abolished both sucrose-inducibility and tuber-specific expression. More extensive deletions that removed the entire repeat had no further effect (Figure 6). In the context of the -920bp promoter these deletions in the promoter proximal repeat abolished neither tuber-specific expression nor sucrose-inducibility. We concluded that the promoter distal repeats were also important in conferring both patterns of expression. We also concluded that the B region of the repeat are of critical importance in mediating the effects of exogenous sucrose and normal developmental cues that activate patatin gene expression in at least tow of the three copies of the repeat.. These results were confirmed by fusing trimers of the entire repeat to -60bp truncated CaMV 35S promoters (essentially a TATA box region that is only capable of basal expression). Figure 7 shows the results of this analysis.

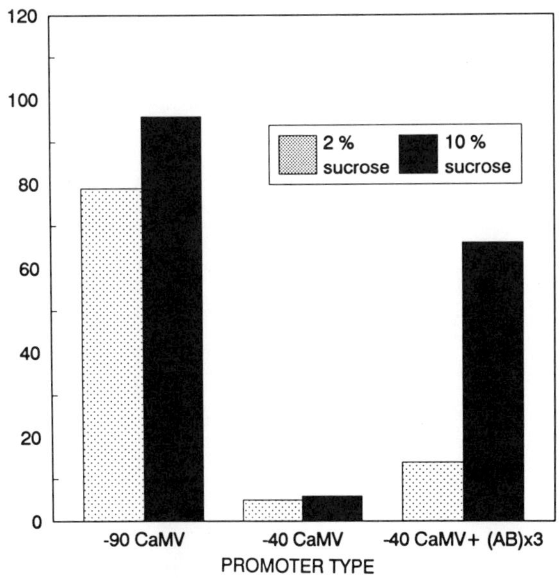

Fig. 7. GUS activity conferred by the A+B repeat fused to a minimal CaMV promoter. Potato plants were transformed with these constructs, together with a control -90CaMV fusion, and resultant plants were assayed for GUS activity after growth on medium containing 2% or 10% sucrose. GUS activity is shown on the vertical axis as pmoles 4-MU/min/μg protein.

The 95mer repeat is able to confer a five-fold level of sucrose-inducibility on the -60CaMV promoter, which in itself is not sucrose-inducible. The -90 CaMV promoter is not significantly inducible either, showing the lack of a general effect of sucrose on transcription levels in these assays. Assays of tuber-specificity of GUS expression conferred by these constructs are still being undertaken. However the levels of expression were very low compared with those obtained from the -360bp construct in sucrose induction experiments. This observation is true for almost all other promoters studied by isolating small regions of DNA and studying their effects individually. These low levels of expression do not negate the findings concerning inducibility and tissue specificity, but rather demonstrate that there is most likely a combination of factors that confers the high levels of expression seen from this promoter in tubers. The *cis* sequences conferring these other effects could be present in the -40 to -400bp region, as these sequences confer high levels of tuber-specific expression on a -90 CaMV 35S promoter (Jefferson et al., 1990). It is intended to extend the survey of protein binding sites to include sequences not already examined in this area, and establish the function of these sequences by deletions and transformation experiments.

EMSA and footprinting experiments have demonstrated the binding of at least two distinct activities to the 95mer repeat that confers sucrose inducibility. One of these activities binds to the AT-rich region, and has been previously noted by Liu et al. (1990). This factor, called TAT-1, has an apparent MW of 35-39kd (data not shown) and binds to the region between -160 and -184bp on the top strand, and -173 and -200bp on the bottom strand, as demonstrated by copper-phenanthroline footprinting (data not shown). This

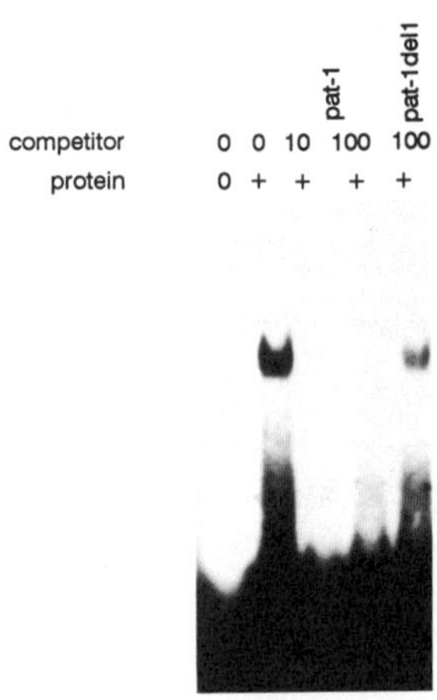

Fig. 8. EMSA experiments using pat-1 as a labelled probe to detect binding of nuclear proteins from developing tubers. Self-competitions with 10- and 100-fold excess of unlabelled pat-1 demonstrate specific binding. Competition with a deleted version of pat-1, pat-1 delA or del1 (shown in Figure 5) shows a reduced affinity of the binding activity for this probe compared to pat-1.

assymetric binding has also been noted in the binding of the NAT-1 and LAT-1 factors to AT-rich sequences in the soybean leghaemoglobin promoter (Jacobsen et al., 1990). It has similarities to HMG proteins, which have been found to bind to AT-rich sequences found in many different plant promoters (Pedersen et al., 1991), and in yeast genes (Struhl, 1985). Not enough consistent information is available to ascribe a function to these sequences in plants, but in yeast they are thought to be non-specific enhancers, and the have been predicted to play a role in nucleosome phasing and attachment to the nuclear scaffold. Another binding activity, demonstrated in Figure 8, which is present in much lower amounts and has a subtantially higher affinity for its binding site(s) than the TAT-1 factor, binds to the distal B region of the repeat (shown as fragment pat-1 in Figure 5). It is present in low levels during early tuber formation and possibly as multiple forms, as several binding activities are consistently eluted from heparin-agarose, and can only be studied after such a partial purification. Smaller regions of the distal region of the repeat also bind the factor, but when this small sequence has a deletion present in its central area equivalent to the deletion from -224 to -208 that abolishes sucrose-responsiveness and tuber-specificity, it no longer binds the factor, and is no longer as effective a competitor for the activity that binds to pat-1. Further studies are being made of this binding activity, as it displays all of the characteristics of a factor that may be involved in conferring these responses. Figure 5 shows that the sequences deleted in repeat B that abolish tuber specificity and sucrose inducibility, and that also abolish binding of the factor(s) found in young developing tubers, are conserved among all three copies of the repeat in the -2400bp promoter. Footprinting experiments and a systematic mutagenesis of pat-1 will delineate the precise binding site(s) on pat-1. This information is necessary for cloning the putative transcription factor(s) involved.

Discussion

Substantial progress has been made in analysing some of the *cis* elements involved in conferring tuber specificity and sucrose-inducibility on the patatin promoter. All of our attention has been directed towards the A and B repeats, and a clear involvement in sucrose-responsiveness has been demonstrated for the B domain. The A domain has many of the characteristics of a general enhancer that binds HMG-like proteins. Its direct role in patatin promoter function is only now being tested. The binding characteristics of the distal versions of the B domain will be investigated, and the effect of specific mutations in these sequences on expression will also be examined. In addition, we intend to begin a systematic study of the -40 to -400bp region in order to determine the binding sites of putative transcription factors, as this region has been shown to confer high levels of tuber-specific expression on a minimal CaMV promoter (Jefferson et al., 1990).

Our preliminary data indicates that one component of the promoter is an element that can respond to exogenous sucrose levels and activate transcription. It is very likely that other components confer a degree of tuber-specificity to expression patterns, but this can be overidden by the sucrose signals, as demonstrated by the high levels of patatin transcription seen in many cell types and tissues after growth on sucrose (Rocha-Sosa et al., 1989; Jefferson et al., 1990). The data obtained for sporamin transcription in tobacco tissue also shows a loss of tuber-root specificity upon the application of high levels of sucrose (Ohta et al., 1991). The question remains as to how important the sucrose-responsiveness of these promoters is in the "normal" induction of tuberisation and patatin transcription. It is relevant to note that sucrose levels of greater that 0.3M can be found in tubers, and this is the same concentration that induces transcription in excised tissues when applied exogenously. It is also relevant that sucrose can induce the transcription of other genes abundantly expressed in tubers, such as GBSS and the S subunit of ADPGPP. A definitive answer to this question can be obtained by investigating other regions of the patatin promoter and other sucrose-responsive promoters, comparing their functions, and examining the effects of combinations of different *cis* sequences on tissue-specificity. Another important question concerns the nature of the stimulus for the sucrose-responsive elements. It is unlikely that sucrose itself is directly involved, as levels of sucrose themselves are not an accurate indicator of the flux through the pathways of carbohydrate metabolism, and it is the flux of photosynthate that is important in determining source-sink strengths (Turgeon, 1989). Perhaps fructose-2,6-bisphosphate is involved in activating the transcription of genes via intermediatary proteins? It's levels are known to respond sensitively to changes in the levels of various 3-carbon and 6-carbon phosphorylated intermediates (reviewed by Stitt, 1990). In yeast and mammalian cells gene regulation by carbohydrates is exemplified by glucose or catabolite repression of genes. This is mediated via a protein kinase in yeast (Celenza and Carlson, 1986) which relieves repression of an invertase gene by phosphorylation of a repressor. Perhaps analogous signal circuits operate in plants, with the added complexity of source-sink variations in carbohydrate levels and the need to regulate these. The isolation of a putative transcription factor binding to the B domain of the patatin promoter will provide an important means of uncovering the mechanisms whereby gene expression in plants is regulated by carbohydrate levels. Once understood, these fundamental regulatory circuits could be altered in order to adapt plants to new uses and environmental conditions. Also, hybrid promoters could be designed for expressing foreign genes at agronomically acceptable levels.

REFERENCES

Arndt, K.T., Styles, C. and Fink, G.R. 1987, Multiple global regulators control *HIS4* transcription in yeast. *Science*, 237: 874-880.

Benfey, P.N. Ren, L. and Chua, N-H. 1990, Tissue-specific expression from CaMV 35S enhancer subdomains in early stages of plant development. *The EMBO Journal*, 9: 1677-1684.

Bevan, M.W., Shufflebottom, D., Edwards, K., Jefferson, J. and Schuch, W. 1989, Tissue- and cell-specific activity of a phenylalanine ammonia-lyase promoter in transgenic plants. *The EMBO Journal*, 8: 1899-1906.

Bevan, M.W. 1991, Gene activity during tuber formation in the potato. In Gene Expression, ed. Grierson, D., (Blackie & Son, London), pp 75-93.

Bevan, M.W. 1984, Binary *Agrobacterium tumefaciens* vectors for plant transformation. *Nucl. Acids Res.*, 12: 8711-8724.

Celenza, J.L. and Carlson, M. 1986, A yeast gene that is essential for release from glucose repression encodes a protein kinase. *Science*, 233: 1175-1180.

Douglas, C.J., Hauffe, K.D., Ites-Morales, M-E., Ellard, M., Paskowski, U., Hahlbrock, K. and Dangl, J.L. 1991, Exonic sequences are required for elicitor and light activation of a plant defense gene, but promoter sequences are sufficient for tissue specific expression. *The EMBO Journal*, 10: 1767-1775.

Frankel, A.D. and Kim, P.S. 1991, Modular structure of transcription factors: implications for gene regulation. *Cell*, 65: 717-719.

Giuliano, G., Pichersky, E., Malik, V.S., Timko, M.P., Scolnik, P.A. and Cashmore, A.R. 1988, An evolutionarily-conserved protein binding sequence upstream of a plant light-regulated gene. *Proc. Natl. Acad. Sci. USA*, 85: 7089-7093.

Hahlbrock, K. and Scheel, D. 1989, Physiology and molecular biology of phenylpropanoid metabolism. *Annu. Rev. Plant Physiol. Plant Mol. Biol.*, 40: 347-369.

Hammer, R.E., Krumlauf, R. Camper, S.A., Brinster, R.L. and Tilghman, S.M. 1987, Diversity of alpha-fetoprotein gene expression in mice is generated by a combination of separate enhancer elements. *Science*, 235: 53-58.

K.D., Paskowski, U., Shulze-Lefert, P., Hahlbrock, K., Dangl, J.L. and Douglas, C.J. 1991, A Parsley 4CL-1 promoter fragment specifies complex expression patterns in transgenic tobacco. *The Plant Cell*, 3: 435-443.

osen, K., Laursen, N.B., Jensen, E.O., Marcker, A., Poulsen, C. and Marcker, K.A. 1990, HMG I-like proteins from leaf and nodule nuclei interact with different AT motifs in soybean nodulin promoters. *The Plant Cell*, 2: 85-94.

Jefferson, R., Goldsborough, A. and Bevan, M.W. 1990, Transcriptional regulation of a patatin-1 gene in potato. *Plant Molecular Biology*, 14: 995-1006.

Jefferson, R.A., Kavanagh, T.A. and Bevan, M.W. 1987, GUS fusions: ß-glucuronidase as a sensitive and versatile gene fusion marker in higher plants. *The EMBO Journal*, 6: 3901-3907.

Johns, E.W., 1982, The HMG chromosomal proteins. E.W. Johns, ed., (London: Academic Press), pp 1-7.

Johnson, R. and Ryan, C.A. 1990, Wound-inducible potato inhibitor II genes: enhancement of expression by sucrose. *Plant Molecular Biology*, 14: 527-536.

Lamb, C.J., Lawton, M.A., Dron, M. and Dixon, R.A. 1989, Signals and transduction mechanisms for activation of plant defenses against microbial attack. *Cell*, 56: 215-224.

Lewin, B. 1990, Commitment and activation at Pol II promoters: A tail of protein-protein interactions. *Cell*, 61: 1161-1164.

Liang, X., Dron, M., Schmidt, J., Dixon, R.A. and Lamb, C.J. 1989, Developmental and environmental regulation of a PAL-GUS gene fusion in transgenic tobacco. *Proc. Natl. Acad. Sci. USA.*, 86: 9284-9288.

Liu, X.J., Prat, S., Willmitzer, L. and Frommer, W.B. 1990, *Cis* regulatory elements directing tuber-specific and sucrose-inducible expression of a chimeric class I patatin promoter/GUS-gene fusion. *Mol. Gen. Genet.*, 223: 201-206.

Lloyd, A., Yancheva, N. and Wasylyk, B. 1991, Transformation suppressor activity of a Jun transcript--ion factor lacking its activation domain. *Nature*, 352: 635-638.

Lois, R., Dietrich, A. Hahlbrock, K. and Schulz, W. 1989, A phenylalanine ammonia-lyase gene from parsley: structure, regulation and identification of elicitor and light responsive *cis*-acting elements. *The EMBO Journal*, 8: 1641-1648.

Mignery, G.A., Pikaard, C.S. and Park, W.D. 1988, Molecular characterization of the patatin multigene family of potato. *Gene*, 62: 27-44.

Ohl, S., Hedrick, S.A., Chory, J. and Lamb, C.J. 1990, Functional properties of a PAL promoter from *Arabidopsis*. *The Plant Cell*, 2: 837-848.

Ohta, S., Hattori, T., Morikami, A. and Nakamura, K. 1991, High-level expression of a sweet potato sporamin gene promoter: ß-glucuronidase (GUS) fusion gene in the stems of transgenic tobacco plants is conferred by multiple cell type-specific regulatory elements. *Mol. Gen. Genet.*, 225: 369-378.

Pedersen, T.J., Arwood, L.J., Spiker, S., Guiltinan, M.J. and Thompson, W.F. 1991, High mobility group chromosomal proteins bind to AT-rich tracts flanking plant genes. *Plant Molecular Biology*, 16: 95-104.

Rocha-Sosa, M., Sonnewald, U., Frommer, W., Stratmann, M., Schell, J. and Willmitzer, L. 1989, Both developmental and metabolic signals activate the promoter of a class I patatin gene. *The EMBO Journal*, 8: 23-31.

Schmid, J., Doerner, P.W., Clouse, S.D., Dixon, R.A. and Lamb, C.J. 1990, Novel and environmental regulation of a bean CHS promoter in transgenic tobacco. *The Plant Cell*, 2: 619-631.

Sheen, J. 1990, Metabolic repression of transcription in higher plants. *The Plant Cell*, 2: 1027-1038.

Staswick, P.E. 1990, Novel regulation of vegetative storage protein genes. *The Plant Cell*, 2: 1-6.

Staswick, P.E., Huang, J-F. and Rhee, Y. 1991, Nitrogen and methyl jasmonate induction of soybean vegetative storage protein genes. *Plant Physiology*, 96: 130-136.

Stitt, M. 1990, Fructose-2,6-bisphosphate as a regulatory molecule in plants. *Annu. Rev. Plant Physiol. Plant Mol. Biol.*, 41: 153-185.

Struhl, K. 1985, Naturally occurring poly (dA - dT) sequences are upstream promoter elements for constitutive transcription in yeast. *Proc. Natl. Acad. Sci. USA*, 82: 1919-8923.

Turgeon, R. 1989, The sink-source transition in leaves. *Annu. Rev. Plant Physiol. Plant Mol. Biol.*, 40: 119-138.

Wenzler, H.C., Mignery, G.A., Fisher, L.M. and Park, W.D. 1989, Analysis of a chimeric class-1-GUS gene in transgenic potato plants: High-level expression in tubers and sucrose-inducible expression in cultured leaf and stem explants. *Plant Molecular Biology*, 12: 41-50.

Weisshaar, B., Armstrong, G.A., Block, A., da Costa e Silver, O. and Halbrock, K. 1990, Light-inducible and constitutively expressed DNA-binding proteins recognizing a plant promoter element with functional relevance in light responsiveness. *The EMBO Journal*, 10: 1777 - 1786.

EXPRESSION OF TRANSFERRED GENES IN TRANSGENIC RICE (TISSUES) AND TOBACCO

L.A.M. Hensgens and R. A. Schilperoort

Leiden University
Clusius Laboratory
Department of Plant Molecular Biology
MOLBAS Research Group
Wassenaarseweg 64
2333 AL Leiden, The Netherlands

INTRODUCTION

During the past ten years genetic engineering of plants has been a central theme in plant molecular biology. Transformation looked promising and without any limitation. Industries participated in research, as it seemed the most promising development for agroindustry since the green revolution. After this initial enthusiasm, industry and university researchers realized that the genetic engineering of plants is still promising but more laborious, slower and more expensive than anticipated. Reasons for this rather skeptic attitude are: - Many crops are difficult to transform; - easily transformed crop are generally economically not interesting; - and the financial investment in this type of research is difficult to earn back. The profits on the major crops (cereals, etc.) are already minimal if not negative and it is difficult to imagine what kind of introduced character would increase the value of these crops enough to pay back the enormous investments. Therefore protein-farming or the manipulation of high value crops like flowers and medicinal plants have become more attractive goals. Transformation gives the opportunity to convert breeding lines into products which can be given trade marks. This simple fact is sometimes regarded as a commercial value being more valuable than any agronomical improvement.

Besides the obstacles of transformation and investment there is the question of what genes are to be introduced. Herbicide resistance will form the subject of public debate. Much research has to be performed to find and/or to characterize genes which upon transfer into plants will convert the plant resistant to diseases caused by insects, viruses, bacteria or nematodes. Resistance against a disease probably requires the transfer and expression of

Gene Conservation and Exploitation
Edited by J.P. Gustafson *et al.*, Plenum Press, New York, 1993

. Many natural variants of plant diseases exist, therefore a ınd precise study of the disease, its natural variants as can be ı the crop and in related crops is required for the genetic of plant resistance. This then might prevent that pests adapt ɔ the introduction of new resistant varieties.

.other problem involves the expression of the transferred genes. The ɹn of the introduced traits if often not as expected. Expression can be ın the greenhouse, but not elsewhere. Expression can be season and ɹn dependent. It has been shown that transferred genes interact with and can inactivate the expression of a previous transferred gene ɩzke et al., 1989). The molecular basis of this inactivation mechanism is ‹nown neither it is known whether it represents a natural process with the .ant to regulate its gene expression. Somoclonal variation is another poorly ınderstood process which influence the phenotypic expression of plant genes. It can be observed in pure races, within single plants and its frequently observed upon *in vitro* culture of plant tissue. Somoclonal variation might represent a fundamental process of the plant to sustain its growth and to adapt itself to the conditions it has to survive before giving rise to an offspring. It must be anticipated that the phenotypic expression of transferred genes will be prone to similar processes. A successful application of genetic engineering therefore requires a fundamental research of gene expression in different species but also in different accessions of the species.

Our knowledge of gene expression in plants is still limited. Only a small number of different genes have been studied under standard laboratory conditions to optimize the reproducibility of the results but limiting our knowledge about their expression in nature under varying conditions.

The conservation of germplasm therefore should not be regarded only as the conservation of a reservoir of pure genetic information but also as a source of information in which this genetic information is expressed in different ways depending on the physical conditions in which the information survived. A parallel can be made with computers the conservation of software (genetic information) will give information and can be run of different types of computers, however when we do not have information on the computer for which the program has been developed it can be anticipated that the programs will have much less value. One can anticipate that for a successful genetic engineering program for plant improvement one has to conserve the alien germplasm. This conservation must be in the normal habitat of the germplasm. This then will enable us in future to isolate and to manipulate genes from the germplasm and to study the phenotypic expression of the genes in the habitat in which they probably have evolved and in which they have been selected.

In this article we want to describe some of our data concerning transformation technologies of rice (*Oryza sativa* L.) and the expression of some transferred genes in rice and tobacco (*Nicotiana tobacum*).

TRANSFORMATION

One of the most difficult points in discussing plant transformation is its definition. In an excellent recent review of the numerous methods of transformation of cereals Potrykus (1990) listed several requested proofs for integrative transformation. One of these, was a tight correlation between physical (Southern blot, *in situ* hybridization) and phenotypic (enzyme assay) data. A second, included a correlation of physiological and phenotypic evidence for transmission to the offspring of the transformed plant. During our transformation and expression studies of tobacco and rice we obtained data showing that some of these points have to be redefined. Phenotypic evidence is obtained as the result of gene expression. Gene expression is the ultimate result of a complex series of processes starting at the gene level and via RNA resulting in the presence of an active protein at the correct location within a plant cell. Table 1 outlines a summary of the processes leading to the ultimate phenotypic expression of a gene.

Many of these processes have hardly been studied. Knowledge has been acquired for only one or few plant species. One of the main scopes of plant

Table 1. Summary of processes and factors determining the ultimate level of gene expression (phenotypic gene expression).

- History of cells (often one copy of a diploid gene is inactive)
- The localization of the gene within the chromosome
- Initiation of transcription
  promoter elements, enhancers, silencers (cell type, stage and hormone dependent)
- Interaction with other genes
- Stability and processing of primary transcript
  editing, splicing, capping, turnover in nucleus etc.
- Export RNA to cytosol
- Stability of mRNA
- Location within the cell of mRNA translation
- Initiation translation
- Translation (progress, fidelity)
- Stability protein, enzyme (turnover)
- Actual enzyme activity in a cell is determined by concentrations of substrate, product, cofactor(s) and inhibitors. The concentrations are influenced by age, type of cell and by environmental factors.
- Studied tissues consist of different specialized cells

transformation is to manipulate and to study the expression of the transferred gene. Therefore, one can not state that transformation is only valid when phenotypic expression can be observed as expected, as did Potrykus (1990).

During transformation only a small number of plant protoplasts or cells are transformed. To economize on labor and to prevent the formation of chimeric plants, a selection step is required. In general, selected based on phenotypic expression of the transferred selection gene is one of the first steps following transformation in many transformation protocols.

For selection, auxotrophy can be used however the use of selective agents such as kanamycin, G418, hygromycin, methotrexate and the amino acid homolog homoarginine is more common. The gene resistant to the selective agent can be either present on the same plasmid as the gene of interest or on a separate plasmid. When both genes are present on two different plasmids they are in 35-65% of the transformed plants expressed. This high frequency of co-transfer suggests that when a cell is transformed more than one DNA molecule is taken up by the cell and becomes integrated into a chromosome.

The selection step following transformation selects those transformed cells which express the resistance gene at that moment. One assumes, therefore that the selection gene is expressed in all cell types and during the whole cell cycle. One might validly wonder whether the transformation of some crops is hampered by this axiom. Most selection genes are driven by the cauliflower mosaic virus (CaMV) 35S promoter which has been successfully used in the transformation of some cereal crops (Fromm et al., 1986; Lyznik et al., 1989; Rhodes et al., 1988; Toriyama et al., 1988; Yang et al., 1988; Zhang et al., 1988; Shimammoto et al., 1989; and Meijer et al., 1991). Apparently these promoter resistance gene combinations are expressed during the period of selection in cells which can regenerate later into plants. It is not known whether cells more prone to regeneration express these combinations at the same moment and level. The selection step with these constructs might therefore counterselect against transformed cells with high regenerative capacity. Studies with other combinations of promoters and selection genes will give more insight in these possibilities.

We have performed molecular biological studies on rice since the end of 1986. Until now work focused on establishing technologies and starting materials essential for genetic engineering of rice: - Cell and tissue cultures, - cryopreservation of cultures, - transformation systems, - gene constructs for transformation and - isolation and characterization of tissue specific and constitutive promoters. For the remainder of the paper we would like to discuss our results as an example of a transformation program. These will highlight some of the mentioned problems concerning expression and definition of transformation. First, the development of different vectors used in the transformation experiments is described. The different transformation technologies applied to rice are discussed together with the problems which we encountered in the (stable) expression of transferred genes.

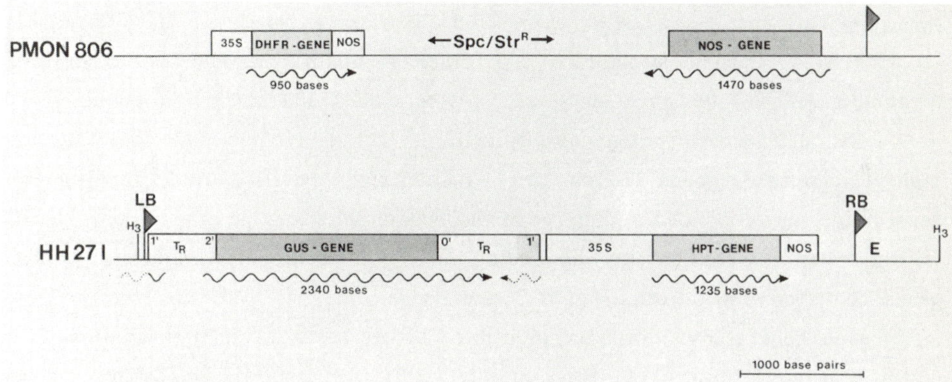

Fig. 1. Plasmids used for the transformation of rice. pMON806 was a gift from Dr. R. Fraley, Monsanto Co., St. Louis, MO, USA. This plasmid contains the methotrexate resistant dihydroxyfolatereductase (*DHFR*) gene under control of the 35S CAMV promoter and nopaline synthase (*NOS*) terminator sequences. It also contains the nopaline synthase gene itself (*NOS*) and the right border of the nopaline Ti-DNA. Between the two genes the bacterial gene coding for streptomycin resistance is present.
HH271 was made in Leiden. It contains the hygromycinephospho-transferase (*HPT*) gene under control of the 35S CAMV promoter and *NOS* terminator sequence and the reporter gene for ß-D-glucuronidase (*GusA*) under control of the promoter of gene 2' and terminator sequence of gene 0'. Gene 2', 0' and 1' are from the TR-DNA of the octopine plasmid pTiAch5. In construct HH10mk (not shown) the *GusA* gene is under control of the gene 1' promoter (box upstream of the *GusA* gene is inverted).
Reading frames and regulatory sequences are indicated by boxes. Direction of transcription and size of transcripts are indicated by the wavy lines. Triangles indicate border sequences involved in the transfer of T-DNA from *Agrobacterium* to plant cells. E indicates sequence which enhances this transfer (Peralta et al., 1986).
For HH271 and HH10mk and HindIII fragment is indicated which can be shuttled between different plasmids.

Vectors and Genes used in Transformation

We found that regenerating rice protoplasts, cell suspensions, callus, germinating seeds and 10 day old rice seedlings were sensitive to hygromycin and methotrexate at rather moderate levels (20 and 0.5 $\mu g.ml^{-1}$ respectively).

Rice was resistant to high levels of the antibiotics kanamycin, G418 and the amino acid homolog homoarginine. Therefore DNA vectors, containing the hygromycin phosphotransferase (*HPT*) gene or the methotrexate resistant dihydroxyfolatereductase (*DHFR*) gene were constructed. These could be used for both direct DNA transformation and *Agrobacterium* mediated transformation. Figure 1 depicts the structures of the constructs and the promoters used.

For the practical use of genetic engineering in rice production it is important to understand gene expression in rice as well as have available strong, tissue specific promoters and other regulatory sequences. Therefore, a number of low or single copy genes were cloned and studied in detail.

By differential colony hybridization cDNA clones corresponding to highly expressed genes which are green-tissue specific, root specific or constitutive were isolated. The genomic counterparts of the single-copy genes were isolated from IR36 and the coding and promoter regions sequenced. The genes contained 2-4 introns (Figure 2 and Table 2).

One gene (*Gos5*) in light inducible. High levels of its transcript can be detected in green tissues (de Pater et al., 1990). Recently it was shown to be homologous to the *PsaH* gene from barley (*Hordeum vulgare* L.) encoding the precursor of a 10.2 Kda photosystem I polypeptide (Okkels et al., 1989).

Extension northern blot analysis showed that the *Gos2* gene is transcriptional active in all tissues studied so far. The region upstream of its putative TATA box contains several binding sites for the tobacco transcription factor TGA-1a (Lam et al., 1989). A nuclear protein from rice, immunological related to TGA-1a, binds to one of these sites (de Pater et al., 1992). Recently it was found that the *Gos2* gene has high homology with a mouse cDNA of unknown function and to the yeast *sui1* gene (A. Schedl, Heidelberg and T. F. Donahue, Indiana University; pers. commun.). The *sui1* gene corresponds to a repressor locus for initiation codon mutants in yeast. It is therefore likely that the *Gos2* gene encodes a protein involved in the initiation of translation. The function of the root specific rice gene *Gos9* remains unknown since no homology with genes present in sequence data banks was found (de Pater and Schilperoot, 1992).

Genes Studied. Expression of the following genes transferred into rice and tobacco was studied:

- The hygromycin phosphotransferase (*HPT*) gene under control of the 35S CAMV promoter and the poly A signal of the *NOS* gene.
- The ß-D-glucuronidase (*GusA*) gene under control of the promoter of the 2' or 2' gene of the T_r region of the *Agrobacterium* tumefaciens octopine Ti plasmid and the poly A signal of the O' T_r gene.

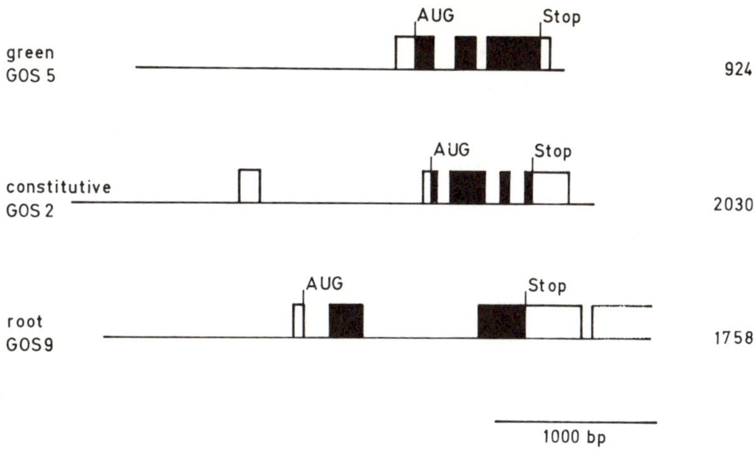

Fig. 2. The structure of the three studied IR36 genes. Sequences coding for mRNA are indicated by boxes. The positions of initiation (AUG) and stop codons are indicated by vertical bars. The black boxes indicate the putative reading frames within exons. Shadowed box in the *GOS9* sequence represents a reading frame not present in the mRNA derived from this gene in roots. The distance between start and end of transcription of the genes is indicated in bp.

- Other *GusA* constructs contained transcriptional and translational fusions with the three rice genes and the poly A signal of the O' T_r gene.
- The dihydrofolate reductase (*DHFR*) gene for methotrexate resistance under control of the 35S CAMV promoter and the *NOS* poly A signal sequence.
- The *NOS* gene with its own regulatory sequences.

The first two genes were present in plamHH plasmids while the *DHFR* gene was present in the pMON806 plasmid (Figure 1). The *GusA* fusions with the rice genes were present in pBluescript vectors (Figure 3).

Table 2. Properties of three IR36 genes. Lengths are given in bases.

	gene 9	gene 5	gene 2
copy number	1-2	1	1
length of the gene	1758	924	2030
number of introns	2	2	4
length of reading frame	474	501	324
length RNA	894	696	692
expression in	root	green tissue	constitutive

Rice was either transformed by direct DNA transformation of protoplasts, both via a chemical procedure and electroporation (Meijer et al., 1991) or by particle bombardment of primary callus. In the last part of this paper expression of transferred genes in rice using alternative and less established methods of transformation are described. Tobacco was transformed by an established leafdisc cocultivation method employing *Agrobacterium* (Horsch et al., 1985).

EXPRESSION OF TRANSFERRED GENES IN DICOTS

Transgenic SR_1 tobacco plants, regenerated from transformed calli and containing 1-2 copies of transferred HH271 or HH10mk DNA (Figure 1), displayed a GUS activity up to 70 $nMOL.min^{-1}.mg^{-1}$ (Hensgens et al., 1991). A detailed quantitative study of GUS activity compared to mRNA levels in independent transgenic tobacco cultures and different parts of F_1 plants showed that no direct correlation existed between the observed steady state levels of GUS activity and mRNA in all studied tissues. In older tissues more GUS activity was found, which is explained by the stable character of the GUS protein and an accumulation of activity over time. Three to ten times higher GUS activities were found in plants grown in vitro than in greenhouse grown sister plants, despite the similar or even higher levels of GUS mRNA in the green house grown plants. Roots from in vitro grown plants displayed three to ten times higher GUS activities than green tissue.

In transgenic tobacco plants containing a translational fusion involving two AUGs in phase, initiation of translocation occurred in leaf material at both AUGs. However, initiation of translation at the first AUG was 10 times more frequent than initiation at the second more distal AUG. In contrast, initiation occurred exclusively at the second AUG in roots from *in vitro* grown plants (Hensgens et al., submitted). These observations clearly demonstrate that the process of (initiation of) translation of mRNA can determine the ultimate level of gene expression.

In transient expression experiments using both tobacco and potato (*Solanum tuberosum* L.) protoplasts the 2' and 1' promoter fusions with the *GusA* reporter gene gave GUS activities 2-3 times higher than those with the 35S CAMV promoter. Expression in stable tobacco transformants was approximately 5-10 times higher than those described with 35S CAMV promoter. The results obtained with the different dicot species show that the fusions between the mannopine synthase genes 2' and 1' and the *GusA* reporter gene gave GUS activities 2-3 times higher than those with the 35S CAMV promoter. Expression in stable tobacco transformants was

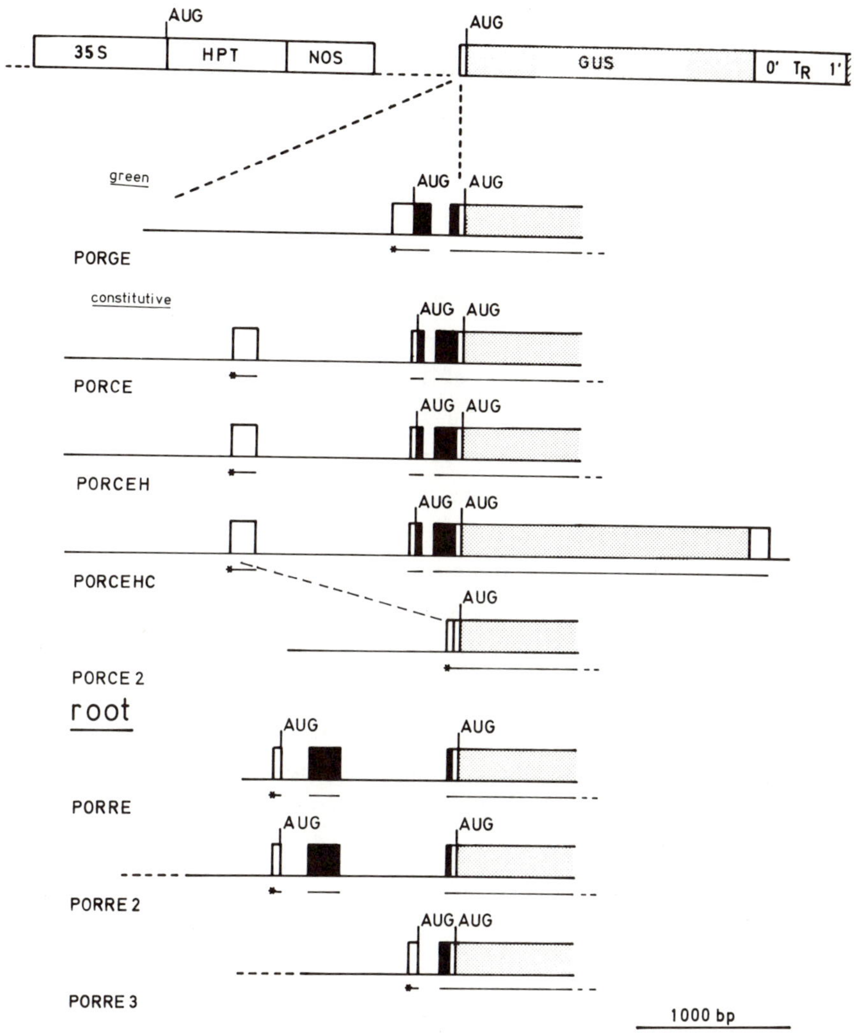

Fig. 3. Fusions between the reportergene *GusA* and three indica IR36 genes. Constructs are made in the polylinker of the plasmid pBluescript. In all constructs except PORCEHC the *GusA* gene is under control of the terminator sequence of gene 0' of the T_R-DNA of the octopine plasmid pTiAch5. In construct PORCEHC the *GusA* readingframe is fused to the last exon and terminator sequence of the constitutive rice gene itself. In all constructs the *HPT* gene under control of the 35S CAMV promoter and *NOS* terminator sequence is present as the selection gene. Sequences of the rice genes found in stable mRNA are indicated by boxes, the rice reading frames by black boxes. Initiation codons of the rice reading frames and *GusA* are indicated (AUG). Putative mRNA sequences are indicated by lines under the constructs. Starts of the mRNAs are indicated by stars. PORCEH and PORCEHC differ from PORCE in the region preceding the AUG of the *GusA* reading frame by a deletion of 12 basepairs to eliminate a possible hairpin in this region.

approximately 5-10 times higher than those described with 35 S CAMV promoter. The results obtained with the different dicot species show that the fusions between the mannopine synthase genes 2' and 1' and the *GusA* reporter gene work very efficiently and give rise to high GUS activities in transformed cells. These results are in contrast with the observations made when the constructs were transferred into rice.

EXPRESSION OF TRANSFERRED GENES IN RICE

Hygromycin or methotrexate resistant rice calli were obtained using the vectors of Figure 1. The transformed state of the tissues as confirmed by Southern blot analysis (Meijer et al., 1991). Despite methotrexate selection for transformed rice cells worked very efficiently we came to the conclusion the use of methotrexate should be omitted as much as possible as selection agent. During the research it became clear that the presence of methotrexate changes the physical properties of the DNA despite the presence of an active *DHFR* gene. Most likely other enzymes involved in the metabolism of nucleotides are inhibited by methotrexate (see also Meijer et al., 1991) leading to incorporation of abnormal nucleotides into the nuclear DNA.

The expression of the transferred genes was studied at different levels. The steady state levels of the transcripts derived from the transferred genes was studied by northern blot analysis. The ultimate phenotypic gene expression was studied by the resistance phenotype and analysis of the produced reporter gene proteins (nopaline synthase or ß-glucuronidase).

Northern Blot Analysis

In transgenic rice calli the observed steady state mRNA levels of both resistance genes (*DHFR* and *HPT*) was high and comparable to those in transgenic tobacco (Figure 4), indicating that the regulatory sequences of both genes work efficiently in the tissues analyzed (calli and suspensions). For the reporter genes *GusA* and *NOS*, however, the level of mRNA was below detection.

Phenotypic Expression of the *GusA* Gene

Despite the presence of the *GusA* gene, GUS activity was not always observed in the transformants. Histochemical staining of callus tissue using X-Gluc (5-bromo-4-chloro-3-indolyl-ß-D-glucuronide) showed that the transformed rice cells contained hardly any GUS activity. Clusters of 10-200 Gus positive cells were observed occasionally in non chimeric transformed callus. However, one third of the suspension cultures initiated from transformed callus tissue displayed GUS activity in a more consistent manner. We observed that GUS activity and number of cells displaying

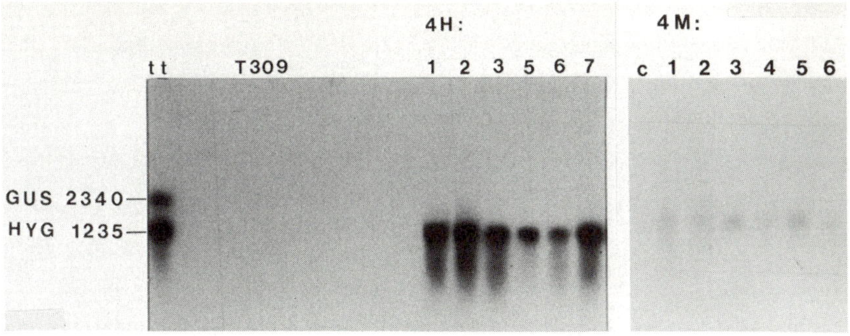

Fig. 4. Transcript analysis of transgenic Hygromycin resistant (4H:) or methotrexate resistant (4M:) rice cell suspensions. RNA was isolated according to Hensgens and Van Os-Ruygrok 1990, 5 μg total RNA was separated on a 1.25% nondenaturing agarose gel, blotted onto GenescreenPlus and hybridized with nicktranslated HH271 plasmid or with a pIC plasmid containing the *DHFR* gene. tt=RNA sample from transgenic tobacco transformed with the HH271 construct. T309 and c:RNA samples from nontransformed T309 suspensions and plants. Sizes of transcripts are indicated in nucleotides.

GUS-activity increased up to approx. 6 days after refreshment of the medium and subsequently disappeared again within 3 days. Both the percentage of cells displaying GUS activity and the level of GUS activity was increased by the addition of 3.10^{-5}M 5-azacitidine to the medium. The presence of GUS activity was also prolonged by this treatment (Meijer et al., 1991). Despite this increase in GUS activity and number of positive cells, no GUS transcript was detectable in cells displaying clear GUS activity (Figure 5).

Comparable findings were reported recently by Harris et al. (1990). They found that transgenic maize (*Zea mays* L.) callus did not stain equally for GUS activity but fluorometric assays with extracts from non-staining callus sectors showed considerable GUS activity. Sectioning of the callus prior to staining did not overcome the presence of such non-staining sectors and it was therefore concluded that negative staining was not due to substrate impermeability. Possibly certain cell types contain inhibitors of the GUS staining reaction which are not active or not present in protein extracts.

It has been stated by de Keijser et al. (1989) that the 2' mannopine promoter present in our construct HH271 is one of the most efficient promoters in transient expression experiments with rice. The maximum activity we found until now in transformed rice (HH271) suspensions is 0.25 $nMol.min^{-1}.mg^{-1}$ of protein extract, which is low compared to that found in transgenic tobacco plants. Tobacco plants containing the same construct, displayed a GUS activity which ranged from 0.1 to 70 $nMol.min^{-1}.mg^{-1}$. However some cell suspensions and calli initiated from the tobacco plants displayed a GUS activity up to 200 times lower than the original plant. Therefore, it appears that the observed reporter gene activity depends on the developmental nature of the cells. The observed non persistent expression of the *GusA* fusions in rice callus and suspensions might be caused by the specific promoters (T_R genes 1' or 2') or terminator (T_R gene 0') used, or factors like translation initiation, compartmentalization of protein synthesis, post translational modification, or position effects within the rice chromosomes. In this context it was interesting that small plantlets regenerated from a transformed embryogenic cell suspension displayed remarkable high levels of ß-D-glucuronidase activity and showed high activity throughout all their tissues especially in the trichomata and stomata (Figure 6). The observed high GUS activity in

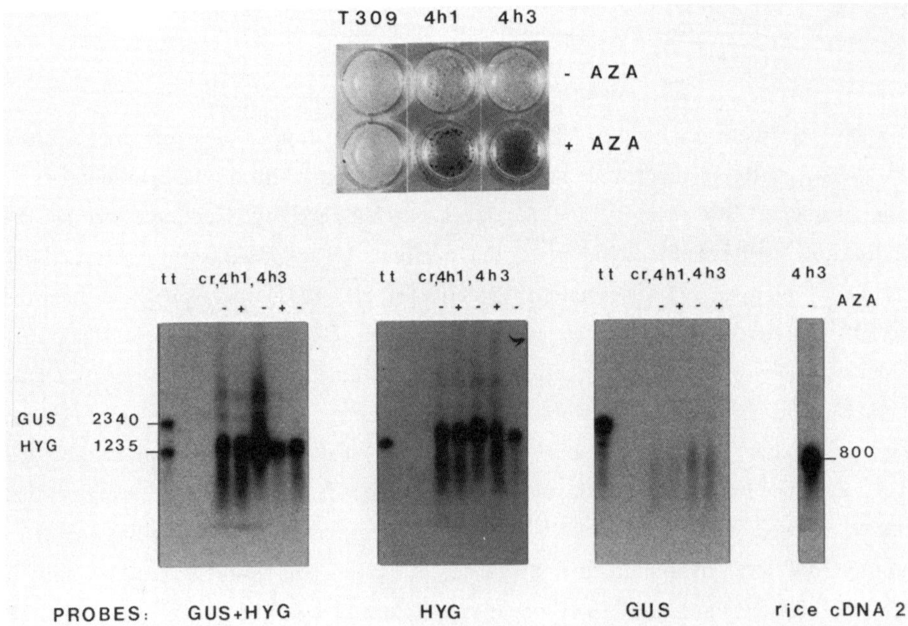

Fig. 5. Histochemical staining for GUS activity in two transformed rice cell suspensions (4h1 and 4h3) grown in the presence of 3.10^{-5}M Azacitidine for 8 days and transcript analysis of the same suspensions. RNA was isolated according to Hensgens and Van Os-Ruygrok 1990, 5 μg total RNA was separated on a 1.25% nondenaturing agarose gel, blotted onto GenescreenPlus and hybridized with nicktranslated HH271 plasmid (probe GUS + HYG) or with isolated fragments corresponding to the *HPT* gene (HYG) or *GusA* gene (GUS). tt = RNA samples from transgenic tobacco transformed with the HH271 construct. cr: RNA samples from nontransformed T309 suspension. A rice cDNA clone corresponding to *GOS2* was used as control to show intactness of isolated RNA. Sizes of transcripts are indicated in nucleotides.

trichomata and stomata of regenerants, indicated that the 2' promoter used may be of potential use for the expression of resistance genes against diseases which are due to infection via these plant structures. Examples of these diseases are Bacterial leaf streak caused by *Xanthomonas campestris*, Narrow brown leaf spot caused by the fungi *Cercospora janseana* (as *c. oryzae*), Minute leaf and grain spot are sometimes caused by species of *Nigrospora* and

Fig. 6. Histochemical assay for GUS activity in leaves from a rice plantlet regenerated from a cell suspension transformed with HH271. Assay was according to Rueb and Hensgens 1990. Different magnifications are shown. Tissue in panel E was partially destained to visualize better the high GUS activity in trichomata and stomata.

Sheath rot caused by *Sarocladium Oryzae* (Sawada). All these organisms infect the rice plant through its stomata, although the latter can also infect via wounds (Ou, 1985). Rice Blast is caused by *Pyricularia oryzae Cavara* (or *P. grisea*) which infects rice mainly by penetration through cuticle and epidermis, although the infection hyphae may also enter through the stomata (Ou, 1985).

Expression of the Nopaline Synthase Gene

Transformed rice tissues containing the pMON806 construct did not contain detectable nopaline synthase activity nor detectable *NOS* mRNA although the presence of intact *NOS* sequences could be determined. Southern blot analysis revealed that two ClaI-sites within the *NOS* gene are almost always methylated. This indicated inactivation by methylation. The addition of 5-azacitidine did not lead, however, to detectable *NOS* products.

Co-Suppression of Transferred Genes

Screening for the expression of transformed traits occurs frequently in the very early stages of genetic engineering experiments. Our observations described above indicate that this occasionally might lead to disposal of transformed material which could have expressed the desired trait upon regeneration at a later stage of growth. In some cases it has been found that in transgenic tobacco the expression of two genes transferred into tobacco are not independent. Often repression of the previously introduced gene can be observed upon introduction and selection of a second resistance gene. The cause of the suppression remains still unclear although methylation of the repressed gene is frequently observed (Matzke et al., 1989; Matzke and Matzke, 1990).

To study the possibility that the expression of the reporter genes *GusA* and *NOS* were hampered by the selection pressure on the expression of the *HPT* or the methotrexate resistant *DHFR* gene, two different experiments were set up. First the methotrexate resistant *DHFR* gene was cloned into HH10mk plasmid. Transgenic rice calli were obtained containing both resistance genes (*HPT* and *DHFR*) or only the *HPT* gene. Transgenic calli were obtained by hygromycin selection and kept under this selection pressure for five months. The calli were divided and grown further with an without selection pressure for two additional months. After this period, both types of calli were tested for Hyg^R, $Meth^R$ and $Hyg^R Meth^R$ phenotypes. The results concluded that upon hygromycin selection, in the first six months of the experiments, the expression of the $Meth^R$ phenotype was lost by some unknown suppression mechanism. The suppression was not released when the hygromycin selection pressure was released. It was further evident that the Hyg^R phenotype is lost in 50% of the calli when the hygromycin selection was eliminated for two months. These results suggested that either the selection genes were integrated into extrachromosomal DNA and became lost upon releasing selection or that by an unknown mechanism the expression of the integrated genes is switched off. Following relaxation of selection pressure the majority of the cells became sensitive. We are currently investigating these possibilities. In this context our successful transformation experiments with immature rice embryos are interesting. These experiments show much higher frequencies of stable expression. In these experiments the transformation and therefore also integration of the transferred DNA into the plant chromosome precedes the callus induction and *in vitro* growth of the transformed tissues. It is tempting to speculate that the *in vitro* growth of the rice cells induces the formation of amplified extrachromosomal DNA as has been found for some rice suspensions (Cuzzoni et al., 1990).

Expression of Homologous Rice Gene - *GusA* Fusions Employing Ballistics

To study the promoter regions of the isolated rice genes, translational and transcriptional fusions with the ß-D-glucuronidase gene were made (Figure 3). First, we focused our attention to the *GusA* fusions with the constitutively expressed rice gene 2. Three rice constructs (PORCE, PORCE2 and PORCEH) were tested and compared with a maize alcoholdehydrogenase gene *GusA* fusion construct (pAI_1GusN; Wang et al., 1988) which was shown to be active in rice cell suspensions (T309, Wang et al., 1988). Rice cell suspensions initiated from the Japonica varieties T309, Nipponbare and the Indica variety IR54 were used as well as a BMS maize cell suspension. In these experiments both stable and transient expression was studied. Constructs were transferred into the different celltypes by bombardment with micro projectiles coated with the different constructs.
In the first particle bombardment experiments it became clear that the two independent initiated T309 suspensions, used for stable protoplast transformation experiments (T4 and T5, Meijer et al., 1991), showed negative results. Also, in previous transient expression experiments where DNA was introduced by chemical or electroporation methods, the suspensions never lead to any transient expression, although, they did give rice to the stable transformants described above.

Using other cell suspensions derived from the varieties Nipponbare and IR54 high GUS activities were observed using the construct PORCE. In this construct the *GusA* readingframe was translational fused within the third exon of the constitutively expressed rice gene *Gos2*. Using this construct, more blue staining was observed in rice suspensions than using the pAI_1GusN construct, containing the maize alcoholdehydrogenase promoter and the first intron of this gene (Figure 7). The latter construct however yielded more intense staining on the homologous BMS maize suspension. These data therefore suggest that the PORCE construct is more active in rice than the maize construct.

The putative constitutive character of the *GusA* construct was also tested by bombardment of different tissues from T309 seedlings. The results of these experiments are shown in the left panel of Figure 8. Gus positive cells were observed in leaf, nodes, root and coleoptile tissues.

In the construct PORCE2 the *GusA* readingframe was positioned immediately after the promoter region of the rice gene 2, therefore translation initiates at the AUG of the *GusA* readingframe. This construct did not yield blue cells except for some light stained cells on inflorescences. It seems that either the presence of the two introns within constructs PORCE and PORCEH are important for high expression or that initiation of translation in the rice sequence present within the 2nd exon proceeds more efficiently than when in the exon 1 *GusA* readingframe fusion of PORCE2. Luehrsen and Walbot (1991) have shown that inclusion of intron 1 of the alcohol dehydrogenase gene in constructs with various reporter genes increases their transient expression in maize cells.

In experiments using *Lilium* bulbscale explants, *Lolium*, and barley the PORCE construct gave rise to blue cells with a staining similar to cells obtained with *GusA* constructs containing the 35S Camv promoter, double enhancer and the AMV virus RNA 4 leader sequence (pMOG18 construct).

The rice gene *GusA* fusions were also stably transferred into rice callus, rice plantlets and tobacco plants employing microparticle bombardment of immature and mature rice embryos. With the PORCE construct many transgenic rice calli were obtained. The expression of the *GusA* fusions was followed by histochemical staining and by measuring GUS activity in protein extracts. GUS activity measuring in extracts from the transgenic rice calli varied from 0 to 1150 $nMOL.mg^{-1}.hour^{-1}$. Transgenic rice calli displayed different staining patterns varying from a homogenous low intensity staining, chimeric staining to very intense staining phenotypes. A large number of these *in vitro* grown calli had the tendency to display less staining upon subculturing. More experimentation is needed to explain these observations.

Regenerated plantlets displayed a very strong staining upon histochemical assaying for GUS activity in the tips of both primary and lateral roots and in young unfolded leaves. Since the *GOS2* gene is supposed to encode a protein involved in the initiation of translation it can be expected to be expressed in cells where ribosomal proteins and other proteins of the translation process are synthesized at high levels. In the roots of the plantlets staining decreased rapidly with distance from the root tip. This indicated that in rice root tissues the GUS enzyme is less stable than in tobacco where GUS activity tended to accumulate in older parts of the plant.

Transgenic tobacco plants containing the rice *GusA* fusion displayed much less staining as it did when present in transgenic rice. Surprisingly, the plants displayed a cell type specific expression close to that of meristematic cells. The rice *GusA* fusion expressed a cell type specificity as in rice, although at a much lower level. Experiments are in progress to study whether the low expression in tobacco was due to the presence of the rice introns or to the rather unique sequence preceding the AUG initiation codon of gene *GOS2*, which clearly deviates from the "normal" sequence found in dicotyledonons (Cavener, 1991; de Pater et al., 1992).

Phenotypic gene expression in transformed rice (tissue) obtained by protoplast transformation of by ballistics has been described in the above

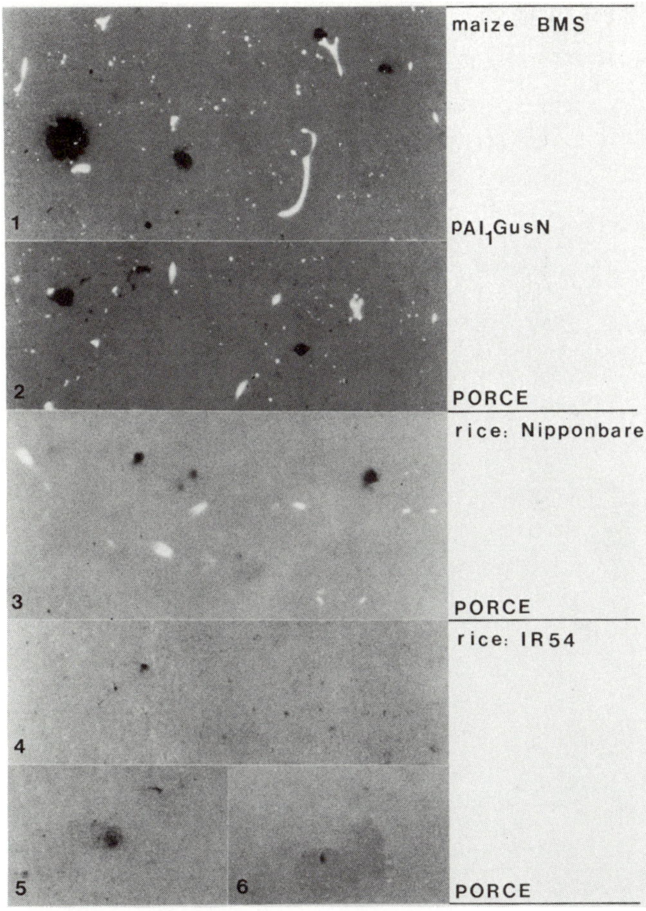

Fig. 7. Transient GUS expression in cell suspensions of rice and maize after microparticle bombardment. Cells were collected on filterpaper discs using funnel and vacuum. Cells were collected 2 days after subculture and plated at a density of $0.6\text{-}1.2 \times 10^6$ cells/filter. Golden microparticles were coated with poly-L-lysine 0.02% and were PEG coated according to Cao et al. (pers. commun.). Prior to every addition, the suspension was sonicated for a short period in a Bransor Ultrasonic cleaner. To 33 mg/ml metal particles 0.1 mg/ml DNA, 8% PEG and 5 mM spermidine-HCl was added. Coating was examined by fluorescence microscopy with 4'-6' diamidino-2-phenyl indole (DAPI). For bombardment 100 μg metal particles were accelerated. After 24 hours at 28°C (for rice) and at 24°C for maize the plant material was tested for GUS activity. Histochemical staining using X-gluc was performed according to Rueb and Hensgens (1989) 2 hours 37°C followed by an overnight incubation at 26°C.

sections. The transformed state of the issue was always confirmed by Southern analysis. It appears that in many cases the phenotypic expression was not stable in time. A similar expression was found in rice (tissue) employing alternative methods of transformation. A description of the methods and the gene expression found is given below and grouped according to the transformation technology employed.

***Agrobacterium* Mediated Transformation**

Most dicotyledonous plants can be transformed by *Agrobacterium* using diverse sources of plant material. Cocultivation of cell suspensions or

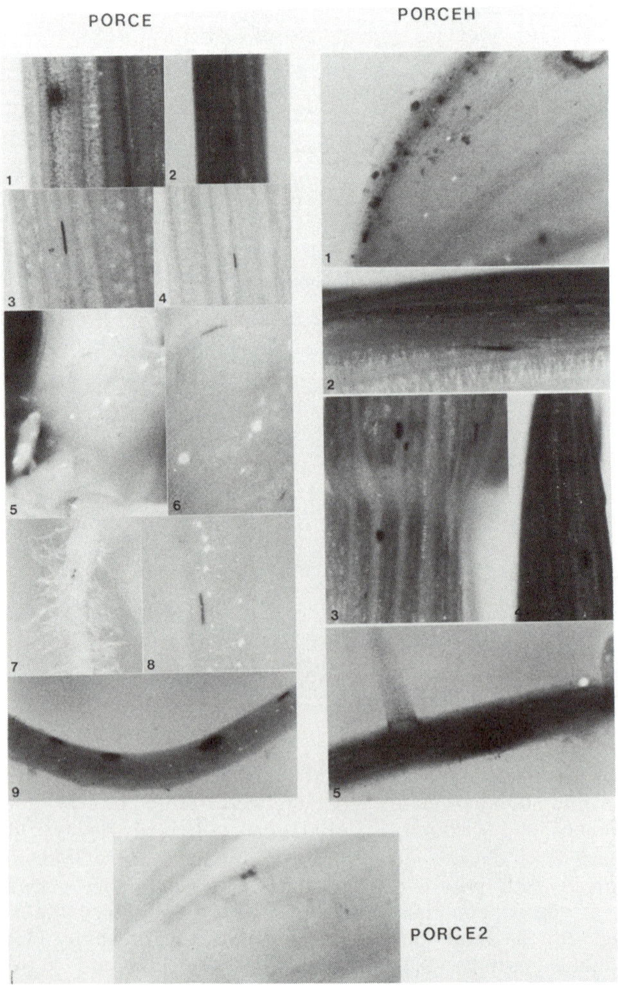

Fig. 8. Transient GUS expression in different tissues of *Oryza sativa* L. Japonica variety T309 plantlets after microparticle bombardment. Leaf, root and coleoptile tissues were from 10 day old seedlings. Inflorescences (right panel 1) were harvested before they emerged from the flag leaf sheath.

protoplasts with *Agrobacterium* is frequently used (Marton et al., 1979; and Van den Elzen et al., 1985). Other methods use leafdiscs (Horsch et al., 1985) or meristematic tissues (Hussey et al., 1989) and have the advantage that protoplasts can be omitted.

Evidence has been published that *Agrobacterium* can also transfer DNA into monocots like *Lilium* (Hooykaas-Van Slogteren et al., 1984), *Gladiolus* (Graves and Goldman, 1987), rice (Raineri et al., 1990) and maize (Grimsley et al., 1986 and 1987). For *Lilium* and *Gladiolus* it was shown by the detection of products produced by T-DNA encoded enzymes (Hooykaas-Van Slogteren et al., 1984; and Graves and Goldman, 1987). For maize and also recently for rice a very sensitive biomonitoring system, called agroinfection, can also be used to detect T-DNA transfer (Grimsley et al., 1987). In agroinfection no selection for transformed cell(s) is possible thus limiting its application in genetic engineering.

It was therefore decided to check possible transformation of rice cells by *Agrobacterium* using more conventional markers: One, changed phenotypes of cell and rice tissues; two, selection markers; three, reporter genes; four, Southern analysis. Several rice tissues were tested, especially the meristematic ones having most regenerative capacity, for susceptibility to transformation by *Agrobacterium*.

Co-Cultivation of Rice Protoplasts. The following experiments were performed. First, partially protoplasted rice cell suspensions were co-cultivated. Several hygromycin resistant calli were obtained, however tests for reporter gene (GUS) activity in these calli were negative. In the co-cultivation experiments the level of apparent false positive (Hyg^R), which escaped hygromycin selection was 10-100 times higher than in the system using direct DNA transformation of protoplasts. Probably the level of hygromycin selection that must be applied to these partially protoplasted rice cells should be much higher.

Co-Cultivation of Rice Inflorescences. Second, promising results were obtained by co-cultivation of immature *Habiganj Aman I* (HAI) young inflorescences with *Agrobacterium* LB4 4404 containing the binary plasmid HH271. This rice variety was chosen for its very fast flowering time (40 days after germination). Unfortunately during the research it appeared to have poor *in vitro* growth characteristics. Panicles were harvested when they reached a size of approx. 3 cm, which occurs 10 days before tillering. Young inflorescences were co-cultivated with acetosyringon induced Agrobacteria (10^8 bacteria/ml; Melchers et al., 1989) for 30 minutes. Thereafter, the inflorescences were washed and laid down on callus inducing agar medium

(Chou et al., 1983). After 2 days, inflorescences were transferred to a medium containing 100 μg/ml cefotaxime and vancomycin (100 μg/ml), to counterselect against *Agrobacterium*, and 10 μg/ml hygromycin. The selection pressure was raised gradually and after 4 months reached a final concentration of 100 μg/ml hygromycin. Twenty-two percent of the inflorescences formed hygromycin resistant calli upon cocultivation. By the histochemical assaying for GUS activity on 85 calli, a variable number of positive cells (Figure 9) was found in 12 of them. The observed activity also varied in time.

In the end, formal proof of transformation should be given by DNA/RNA analysis. Unfortunately, the variety used (HAI) did not easily form suspension cultures from calli. Ultimately three badly growing suspensions were obtained on nonselective media. DNA/RNA analysis of this material showed that only one of the 3 lines contained DNA derived from HH271.Southern analysis using the enzymes XbaI, NcoI and NruI did not yield distinct bands but only smears of DNA which were not detected in controls using nontransformed rice tissues. No GUS transcript could be detected.

Using inflorescences of the rice variety *Japonica* T309 and the *Agrobacterium* strain LBA4404 carrying the construct HH271 a large number of hygromycin resistant calli were obtained which survived selection on media containing 200 μg/ml hygromycin. Nine weeks after the start of the experiment approx. 30% of the tested calli contained GUS activity upon a histochemical test for GUS activity. However, the observed staining was not uniform. Fourteen weeks later positive cells were found in only one of the 300 tested calli. At this moment we can only speculate about the loss of detectable GUS activity in these calli: One, copynumber of the transferred genes was low and therefore the GUS activity was around the threshold level of detection by histochemical staining; two, the *GusA* gene for which no selection was performed over time became inactivated; three, the increased selection pressure on the expression of the hygromycin phosphotransferase gene during the course of the experiment (the hygromycin concentration was gradually increased from 20 to 200 μg/ml) might interfere with the expression of the *GusA* gene.

Rice Seedling Inoculation. Inoculation of young 6-7 day old seedlings (2 cm high) with different *Agrobacterium* strains was performed to ascertain whether or not the young meristematic tissue present within the rice coleoptiles could be used as a target for *Agrobacterium* mediated transforation and subsequent regeneration of transformed cells. A preliminary experiment used the deepwater variety HAI seedlings, the *Agrobacterium rhizogenes* strain LBA9402, and wounds made close to the coleorhiza just above the meristematic zone. Roots with a phenotype similar to "hairy roots" developed from the wound. The experiment was repeated with *Japonica* T309 seedlings. Strong necrotic responses of the tissue surrounding the inoculation site were observed within 10 days after inoculation.Such hypersensitive responses certainly have a negative effect on the growth of any transformed cell. Therefore we studied in detail the response of T309 seedlings following inoculation with various *Agrobacterium* strains to see whether roots or callus could be induced while necrosis could be prevented.

Agrobacteria with an without different wild type Ti- and Ri-plasmids were used. Also binary vector systems were tested using different helper Ti-or Ri-derived plasmids, which contained all virulence genes but no T-DNA, together with a second plasmid carrying only T-DNA but no virulence genes. In addition, several mutants in specific T-DNA genes or virulence genes were screened for their effect on Japonica T309 seedlings. Ten days after inoculation 10 plantlets per strain were visually inspected and sections of the inoculated tissue were examined. Some examples are shown in Figure 10. The main conclusions were as follows:

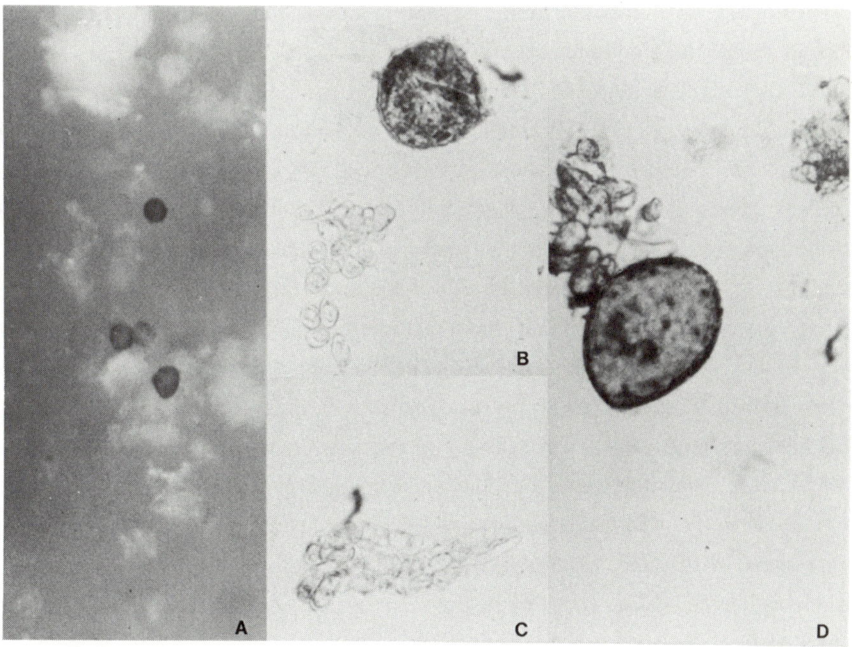

Fig. 9. Histochemical assay for GUS activity in hygromycin resistant rice calli obtained by co-cultivation of Habiganj Aman II inflorescences on callus inducing medium. *Agrobacterium* LBA4404 containing the HH271 construct on a wide host range plasmid was used for co-cultivation. Assay was according to Rueb and Hensgens (1990). Different magnifications of different cells are shown. Note the very heterogeneous activity.

1) Wild type *Agrobacterium tumefaciens* strains gave rise to severe necrosis, while induction of the virulence genes by preincubation with acetosyringon made necrosis even more severe. An exception was the *Agrobacterium rhizogenes* strain LBA 9402, which gave a moderate necrotic response. On *Japonica* T309 seedlings, however, no formation of "hairy roots" was observed.

2) A mutation in one of the virulence genes int he presence of an intact wild type T-DNA region abolished any necrotic response. Also, *Agrobacterium* strains with a mutation in gene 1 or 2, coding for the enzymes IAM synthase and IAA synthase respectively (Van Onckelen et al., 1986; Schröder et al., 1984), did not induce necrosis on the rice seedlings.

3) LBA4460, an *Agrobacterium* strain with a helper plasmid carrying the virulence genes, together with the T_R-DNA of *Agrobacterium rhizogenes* 1855 on a binary plasmid gave a necrotic response (Figure 9). On the T_R-DNA the genes homologous to the genes 1 and 2 of the T-DNA of Ti plasmids were present (Offringa et al., 1986).

All these data together indicate that necrosis induced by *Agrobacterium* is dependent on the combined expression of both the virulence genes and genes 1 and 2 for auxin synthesis. Such a situation strongly suggests that necrosis is due to the transfer of T-DNA into the rice cells and the expression of genes 1 and 2 in these cells. It has been demonstrated (Hooykaas, pers. commun.) that transfer of genes 1 and 2 into stem cells of the monocot *Narcissus* leads to a swelling due to cell elongation, showing that these genes are expressed in monocots. It can be speculated that the necrotic response in rice is the result of auxin synthesis in cells that have acquired T-DNA. Also Lowe and Krul (1991) found an auxin gene (tryptophan monooxygenase) dependent necrosis on stem sections of different *Vitus* species. Until now no data have been published showing that the auxin genes are expressed in Agrobacteria. Therefore, the data indicated that the induced necrosis was due to the transfer of T-DNA into the rice cells and their expression in these cells. Our results are in agreement with the results of Raineri et al. (1990), who showed that germinating rice embryos wounded in the scutellum region can be transformed by *Agrobacterium* and that this transformation was dependent on Ti-plasmid encoded virulence genes. They also showed that a necrotic or hypersensitive response of the inoculated embryo was dependent on the rice cultivar used. No regenerants were obtained, probably because of the bad regenerative characteristics of the inoculated tissues.

Although genetic engineering of rice is already possible by transformation of rice protoplasts, the potential use of the *Agrobacterium*

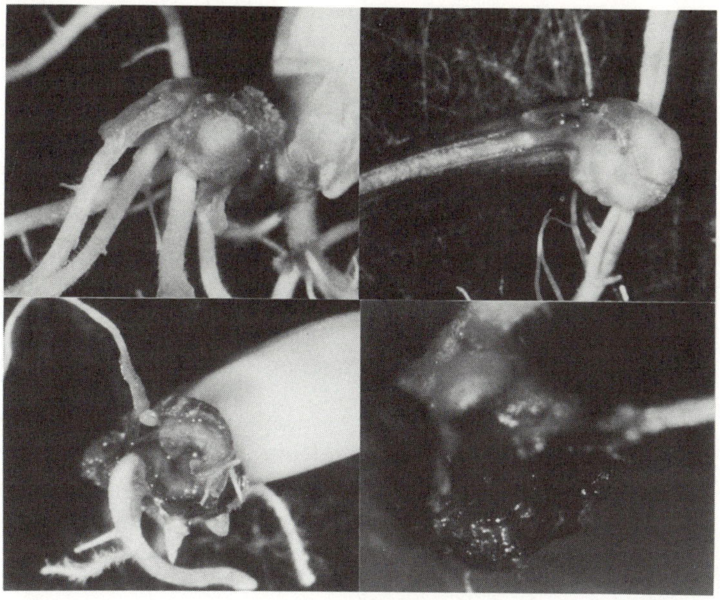

Fig. 10. Necrosis on wounded *Oryza japonica* T309 seedlings inoculated with *Agrobacterium* LBA4404 (T-DNA deleted), LBA4463 (LBA4404 containing the genes 6a and 6b from the T_L-DNA cloned on a wide host range plasmid within borders of transfer) and LBA4460 (LBA4404 containing the T_R-region of the Ri plasmid 1855 on a wide host range plasmid). In left side bottom photograph also a wide host range plasmid carrying the HH271 construct was present within LBA4460. 10 days after wounding and inoculation transsections were examined and photographed. Some representative images are shown.

system is worth further evaluation, since it might lack the disadvantages of the protoplast system. Rice protoplast transformation is complicated and time consuming in that it requires the induction and isolation of cell suspensions for each variety to be transformed. In addition, regeneration of transformed cells into flowering and fertile plants is laborious, time consuming and still unpredictable. The recent development of transformation technology mediated by microparticles however probably forms a better alternative for monocotyledon plant transformation. The technique has recently been shown to be very efficiently for rice in the laboratory of the author. Very high transformation frequencies were obtained using embryos from immature seed. From 240 bombarded embryos more than 350 transgenic plants were obtained. After bombardment embryos were laid down on a callus induction medium and three days later a high selection pressure for the presence of the HPT gene was applied. In total 45 independent hygromycin resistant callus displaying GUS activity regenerated into fertile plants. It was observed that the expression in these transgenic plants and tissues was more consistent as the expression in transgenic material obtained with stable rice suspensions. One can imagine that in the transformation experiment with the embryos the tissues were exposed for a short period to high levels of 2,4-D an artificial hormone used to induce the *in vitro* growth of the rice cells. This hormone is suspected to have many side effects including mutagenic activity. Since the embryo tissue is the tissue with the best regenerative capacity of the rice plant and callus that regenerate to plants can be obtained from most rice varieties rather easily it can be anticipated that the ballistics employing embryos from immature rice seed will work for most varieties. The high frequency in which independent transformed plants can be obtained forms another advantage. The price of the apparatus, the macro-projectiles and an eventual license policy however will form a threshold for its application in developing countries.

Pollen Tube Pathway Transformation

Since the paper of Luo and Wu (1988) described the pollen tube procedure for the transformation of rice as rather simple and efficient, we have repeated their experiment to compare it with other methods.

We used the rice variety *Japonica* T309 together with the construct HH271 (*GusA*; *HPT*) in the plasmid pIC-20H (Figure 1). F_0 and F_1 plants were screened for GUS activity and approx. 20% of the progeny displayed GUS activity above background found in non transgenic T309 plants. Some progeny displayed a transient but no clear hygromycin resistance when F_1 seeds were germinated in liquid medium containing 50 μg hygromycin/ml. Based on this resistance, DNA was isolated and analyzed from 12 parental F_0 plants. Two of them displayed a weak smeared high molecular weight signal upon Southern analysis.

Nine F1 plants derived from one of these (PTP42) were analyzed by Southern analysis of DNA isolated from pooled leaves. A 3190 bp band corresponding to the *GusA* XbaI fragment was observed in this analysis. Another high molecular weight band although less intense was also observed. The intensities of the two bands suggested a copy number lower than 0.4 copies per diploid genome. To determine whether this low intensity was due to a high number of untransformed leaves present in the pooled leaves, Southern analysis was performed on tissues isolated from individual F_1 plants. Since in GUS assays high activity was observed in the nodal regions of these F_1 plants, DNA isolation was combined with RNA isolation from these nodal regions to analyze transcripts in these tissues. Unexpected, Southern analysis on DNA of these nodal regions displayed a completely different pattern than that found for the pooled leaves. All selected individual F_1 plants were positive and displayed a strong hybridizing band present in approx. 1 copy per diploid genome. This strong hybridizing band corresponded to the *HPT* gene

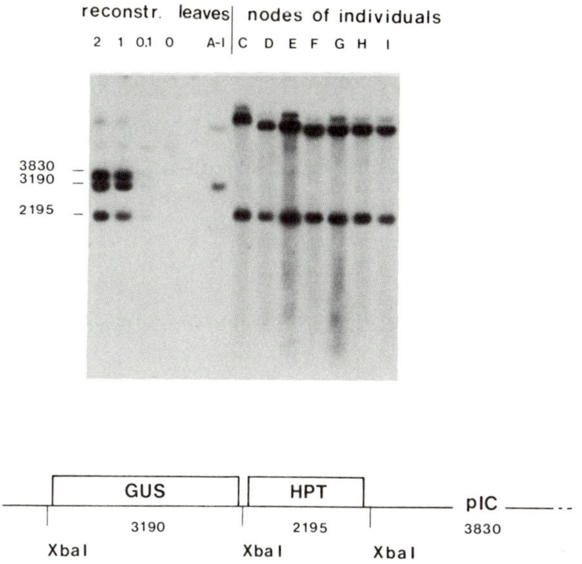

Fig. 11. Southern blot analysis of DNA isolated from pooled leaves from nine plants A-1 and nodes from some of the individual F_1 rice plants obtained by transformation via the pollen tube pathway. DNA was isolated according to Hensgens and Van Os-Ruygrok (1990). 7 μg DNA was digested with Xbal, separated on a 0.8% agarose, 1xTBE gel and transferred to GenescreenPlus and hybridized with nicktranslated HH271 DNA. reconstr.: Reconstruction in which 7 μg of control T309 DNA was digested with XbaI and mixed with 2, 1 or 0, 1 copies of HH271 DNA per diploid T309 genome. The HH271 DNA was isolated from a *dam* *E. coli* strain and digested with XbaI.
Two hybridizing bands are visible in pooled DNA. The 3190 bp band corresponds to the XbaI fragment containing the complete *GusA* gene (see also figure 3). Note that plasmid DNA HH271 used in the transformation was methylated by the *dam* methylation system of *E. coli* on the first two Xbal sites.

fragment (Figure 11). In the pooled leaves DNA of the nine F_1 plants however a 3190 bp band corresponding to the *GusA* gene was observed. We hypothesize that the observed differences are caused by a different and specific methylation of the XbaI sites in leaves and nodes, while the methylation becomes more frequent in the older parts of the leaves. To further substantiate the integration of the vector DNA into rice chromosome, F_2 seeds were analyzed by *in situ* hybridization analysis. A low but significant number of cells displaying a hybridization site on a rice chromosome were detected thus indicating DNA integration (J. P. Gustafson, pers. commun.).

Northern analysis of RNA isolated from the nodal regions of plants transformed by the pollen tubepathway revealed the presence of a transcript derived from the used construct. The size of this transcript however did not correspond to the expected size of the *HPT* or *GusA* gene. The exact nature of the transcript has still to be revealed by PCR analysis.

Despite the fact that in the obtained transformed plant expression of the transferred genes was not as expected, our data suggested that the pollen tube procedure is capable of being useful for transformation of rice. However, considering its laborious nature, another selection marker than hygromycin resistance should be considered. For example, a herbicide resistance gene could be used permitting direct selection of stable transformants expressing the transferred genes in the F_1 generation.

From the results discussed in this paper it is clear that relatively little is known of the expression of transferred genes in monocots and in plants in general and it can be anticipated that these studies will reveal many surprises in future. Many factors like growth conditions, genetic background of the cells, integration place and cell type have influence on their expression. When more knowledge has been acquired for these factors it can be expected that our knowledge and understanding of the expression of plant genes and transferred genes will increase in parallel.

ACKNOWLEDGEMENTS

We should like to thank S. Rueb, Ellen van Os-Ruygrok, L. Bakker and Bram Wetselaar for their skillful technical assistance and B. Nutter and F. van der Mark for their help in some experiments. This research was partly funded by a grant of the Rockefeller Foundation.

REFERENCES

Benfey, P. N., and Chua, N.-H., 1989, Regulated genes in transgenic plants, *Science* 244:174-181.

Cavener, D. R., and Ray, S. C., 1991, Eukaryotic start and stop translation sites, *Nucl. Acids. Res.* 19:3185-3192.

Chou, K. T., Ge, K. L., Tsai, I.S., Yang, C. S., and Yang, H. W., 1983, Callus induction and redifferentiation of different hybrid rice plants, *in*: "Cell and tissue culture techniques for cereal crop improvement," pp. 207-213, Science Press, Beijing, China, IRRI, Manila, Philippines.

Chu, C. C., Wang, C. C., Sun, C. S., Hsu, C., Yin, K. C., Chu, C. Y., and Bi, F. Y., 1975, Establishment of an efficient medium for anther culture of rice through comparative experiments on the nitrogen sources, *Sci. Sinica* 18:659-668.

Cuzzoni, E., Ferretti, L., Giordani, C., Castiglione, S., and Sala, F., 1990, A repeated chromosomal DNA sequence is amplified as a circular extrachromosomal molecule in rice (*Oryza sativa* L.), *Mol. Gen. Genet.* 222:58-64.

Dehayes, A., Herrera-Estrella, L., and Caboche, M., 1985, Liposome mediated transformation of tobacco mesophyll protoplasts by an *Escherichia coli* plasmid, *EMBO J.* 4:2731-2735.

Dekeyser, R., Claes, B., Marichal, M., Van Montagu, M., and Caplan, A., 1989, Evaluation of selectable markers for rice transformation, *Plant Physiol.* 90:217-223.

De la Pena, A., Lörz, H. and Schell, J., 1987, Transgenic rye plants obtained by injecting DNA into young floral tillers, *Nature* 325:274-276.

Frearson, E. M., Power, J. B., and Cocking, E. C., 1973, The isolation, culture and regeneration of *Petunia* leaf protoplasts, *Developmental Biology* 33:130-137.

Fromm, M. E., Taylor, L. P., and Walbot, V., 1986, Stable transformation of maize after gene transfer by electroporation, *Nature* 319:791-793.

Graves, A. C., and Goldman, S. L., 1987, *Agrobacterium tumefaciens* mediated transformation of the monocot genus *Gladiolus*: detection of the expression of T-DNA-encoded genes, *J. Bacteriol.* 169:1745-1746.

Grimsley, N., Hohn, B., Hohn, T., and Walden, R., 1986, Agroinfection, an alternative route for viral infection of plants using the Ti plasmid, *Proc. Natl. Acad. Sci. U.S.A.* 83:3282-3286.

Grimsley, N., Hohn, T., Davies, J. W., and Hohn, B., 1987, *Agrobacterium* mediated delivery of infectious maize streak virus into maize plants, *Nature* 325:177-179.

Harris, R. R., DeRobertis, G. A., Pierce, D. A., Moynihan, M. R., and Everett, N. P., 1990, Heterogeneity of X-gluc staining in transgenic maize callus, Abstract A5-22, Abstracts VIIth Intl. Congr. on Plant Tissue and Cell Cult., p. 176, Amsterdam.

Hensgens, L.A.M., and Van Os-Ruygrok, P. E., 1989, Isolation of RNA and DNA from different rice tissues, *Rice Genetic Newsletters* 6:163-168.

Hensgens, L.A.M., and Van Os-Ruygrok, P. E., 1990, The design of gene constructs for transformation of rice and grass, *in*: "Agricultural biotechnology in focus in the Netherlands," Dekkers, J. J., van der Plas, H. C. and Vuijk, D. H., eds., PUDOC, Wageningen, 25-30.

Hensgens, L.A.M., Meijer, E.G.M., Van Os-Ruygrok, P. E., Rueb, S., de Pater, B.S., and Schilperoort, R. A., 1991, Rice transformation, vectors and expression of transferred genes, *in*: "Rice Genetics II", International Rice Research Institute, Manila, Philippines, 585-596.

Hensgens, L.A.M., Fornerrood, M.W.J., Rueb, S., Winkler, A.A.,v.d. Steen, S. and Schilperoort, R. A., 1991, Translation controls the expression level of a chimeric reporter gene, *Plant Mol. Biol.*, submitted.

Hooykaas-Van Slogteren, G.M.S., Hooykaas, P.J.J., and Schilperoort, R. A., 1984, Expression of Ti-plasmid genes in monocotyledonous plants infected with *Agrobacterium tumefaciens*, *Nature* 311:763-764.

Horsch, R. B., Fry, J. E., Hoffmann, N. L., Eichholz, D., Rogers, S. G., and Fraley, R. T., 1985, A simple and general method for transferring genes into plants, *Science* 227:1229-1231.

Hussey, G., Johnson, R. D., and Warren, S., 1989, Transformation of meristemic cells in the shoot apex of cultured pea shoots by *Agrobacterium tumefaciens* and *A. rhizogenes*, *Protoplasma* 148:101-105.

Meijer, E.G.M., van Iren, F., Schrijnemakers, E., Hensgens, L.A.M., van Zijderveld, M., and Schilperoort, R. A., 1991, Retention of the capacity to produce plants from protoplasts in cryopreserved cell lines of rice (*Oryza sativa* L.). *Plant Cell Reports*, 10:171-174.

Jefferson, R. A., Kavanagh, T. A., and Bevan, M. W., 1987, *GUS* fusions: ß-glucuronidase as a sensitive and versatile gene fusion marker in higher plants, *EMBO Journal* 6:3901-3907.

Kao, K. N., 1977, Chromosomal behaviour in somatic hybrids of Soybean - *Nicotiana glauca*, *Molecular and General Genetics* 150:225-230.

Krens, F. A., Molendijk, L., Wullems, G. J., and Schilperoort, R. A., 1982, *In vitro* transformation of plant protoplasts with Ti-plasmid DNA, *Nature* 296:72-74.

Lam, E., Katagiri, F., and Chua, N.-H., 1990, Plant nuclear factor ASF-1 binds to an essential region of the nopaline synthase promoter, *J. Biol. Chem.* 265:9909-9913.

Linsmaier, E. M., and Skoog, F., 1965, Organic growth factor requirements of tobacco tissue cultures, *Physiol. Plant.* 18:100-127.

Lowe, B. A., and Krul, W. R., 1991, Physical, chemical, developmental, and genetic factors that modulate the *Agrobacterium*-Vitis interaction, *Plant Physiol.* 96:121-129.

Luehrsen, K. R., and Walbot, V., 1991, Intron enhancement of gene expression and the splicing efficiency in maize cells, *Mol. Gen. Genet.* 225:81-93.

Luo, Z.-x., and Wu, R., 1988, A simple method for the transformation of rice via the pollen-tube pathway, *Plant Molecular Biology Reporter* 6:165-174.

Lyznik, L. A., Ryan, R. D., Ritchie, S. W., and Hodges, T. K., 1989, Stable co-transformation of maize protoplasts with the *gusA* and *neo* genes, *Plant Mol. Biol.* 13:151-161.

Man-Si Wang, Zapata, F. J., and de Castro, D. C., 1987, Plant regeneration through somatic embryogenesis from mature seed and young inflorescence of wild rice (*Oryza perennis Moench*), *Plant Cell Reports* 6:294-296.

Marton, L., Wullems, G. J., Molendijk, L., and Schilperoort, R. A., 1979, *In vitro* transformation of cultured cells from *Nicotiana tabacum* by *Agrobacterium tumefaciens*, *Nature* 227:129-131.

Matzke, M. A., Primig, M., Tmovsky, J., and Matzke, A.J.M., 1989, Reversible methylation and inactivation of marker genes in sequentially transformed tobacco plants, *The EMBO Journal* 8:643-649.

Matzke, M. A., and Matzke, A.J.M., 1990, Gene interactions and epigenetic variation in transgenic plants, *Development Genetics* 11:214-223.

Meijer, E.G.M., Schilperoort, R. A., Rueb, S., Zijderveld, M., Van Os-Ruygrok, P. E., and Hensgens, L.A.M., 1991, Transgenic rice cell lines and plants: Expression of transferred chimeric genes, *Plant Mol. Biol.*, 16:807-820.

Melchers, L. S., Regensburg-Tuïnk, A.J.G., Schilperoort, R. A., and Hooykaas, P.J.J., 1989, Specificity of signal molecules in the activation of *Agrobacterium* virulence gene expression, *Molecular Microbiology* 3:969-977.

Müller, A. J., and Grafe, R., 1978, Isolation and characterisation of cell lines of *Nicotiana tabacum* lacking nitrate reductase, *Mol. Gen. Genet.* 161:67-76.

Murashige, T., and Skoog, F., 1962, A revised medium for rapid growth and bioassays with tobacco tissue cultures, *Physiol. Plant.* 15:473-497.

Offringa, I. A., Melchers, L. S., Regensburg-Tuïnk, A.J.G., Costantino, P., Schilperoort, R. A., and Hooykaas, P.J.J., 1986, Complementation of *Agrobacterium tumefaciens* Ti *aux* mutants by genes from the T_R-region of the Ri plasmid of *Agrobacterium rhizogenes*, *Proc. Natl. Acad. Sci. USA* 83:6935-6939.

Ou, S. H., 1985, "Rice Diseases", The Cambrian News (Aberystwyth) Ltd., Great Britain, second edition.

de Pater, B. S., Hensgens, L.A.M., and Schilperoort, R. A., 1990, Structure and expression of a light-inducible shoot-specific rice gene, *Plant Mol. Biol.* 15:399-406.

de Pater, B. S., van der Mark, F., Katagiri, F., Rueb, S., Schedl, A., Chua, N.-H., Schilperoort, R. A., and Hensgens, L.A.M., The promoter of the rice gene GOS2 is active in various different monocot tissues and binds rice neclear factor ASF-1. *Plant Journal*, submitted.

de Pater, B. S. and Schilperoort, R. A., 1992. Structure and expression of a root specific rice gene, Plant Mol. Biol. 18:161-164.

Peralta, E. G., Hellmiss, R., and Ream, W., 1986, Overdrive: A T-DNA transmission enhancer on the *Agrobacterium tumefaciens* tumour inducing plasmid, *EMBO J.* 5:1137-1142.

Potrykus, I., 1990, Gene transfer to cereals: An assessment, *Bio/Technology* 8:535-542.

Raineri, D. M., Bottino, P., Gordon, M. P., and Nester, E. W., 1990, *Agrobacterium*-mediated transformation of rice (*Oryza sativa* L.), *Biotechnology* 8:535-542.

Rhodes, C. A., Pierce, D. A., Mettler, I. J., Mascarenhas, D., and Dettmer, J. J., 1988, Genetically transformed maize plants from protoplasts, *Science* 240:204-207.

Rueb, S., and Hensgens, L.A.M., 1989, Improved histochemical staining for ß-D-glucuronidase activity in monocotyledonous plants, *Rice Genetic Newsletters* 6:168-169.

Schröder, G., Waffenschmidt, S., Weiler, E. W., and Schröder, J., 1984, The T-region of Ti plasmids codes for an enzyme synthesizing indole-3-acetic acid, *Eur. J. Biochem.* 138:387-391.

Schulze-Lefert, P., Dangl, J. L., Becker-André, M., Hahlbrock, K., and Schulz, W., 1989, Inducible *in vivo* DNA footprints define sequences necessary for UV light activation of the parsley chalcone synthase gene, *EMBO J.* 8:651-656.

Toriyama, K., Arimoto, Y., Uchimiya, H., and Hinata, K., 1988, Transgenic rice plants after direct gene transfer into protoplasts, *Bio/technology* 6:1072-1074.

Van der Elzen, P., Lee, K. Y., Towsend, J., and Bedbrook, J., 1985, Simple binary vectors for DNA transfer to plant cells, *Plant Molecular Biology* 5:149-154.

Van Onckelen, H., Prinsen, E., Inzé, D., Rüdelsheim, P., Van Lijsebettens, M., Follin, A., Schell, J., Van Montagu, M., and De Greef, J., 1986, *Agrobacterium* T-DNA gene 1 codes for tryptophan 2-monooxygenase activity in tobacco crown gall cells, *FEBS Letters* 198:357-360.

Wang, M.-S., Zapata, F. J., and de Castro, D. C., 1987, Plant regeneration through somatic embryogenesis from mature seed and young inflorescence of wild rice (*Oryza perennis Moench*), *Plant Cell Reports* 6:294-296.

Wang, C. C., Klein, T. M., Fromm, M., Cao, J., Sanford, J. C., and Wu, R., 1987, Transient expression of foreign genes in rice, wheat and soybean cells following particle bombardment, *Plant Molecular Biology* 11:433-439.

Yang, H., Zhang, H. M., Davey, M. R., Mulligan, B. J., and Cocking, E. C., 1988, Production of kanamycin resistant rice tissue following DNA uptake into protoplasts, *Plant Cell Rep.* 7:412-425.

Zhang, E. M., Yang, H., Rech, E. L., Golds, T. J., Davis, A. S., Milligan, B. J., Cocking, E. C., and Davey, M. R., 1988, Transgenic rice plants produced by electroporation-mediated plasmid uptake into protoplasts, *Plant Cell Rep.* 7:379-384.

ENHANCEMENT OF PLANT PRODUCTIVITY BY MANIPULATION OF ADPGLUCOSE PYROPHOSPHORYLASE

Thomas W. Okita,[1] Paul A. Nakata,[1] Kathryn Ball,[2]
Brian J. Smith-White,[2] and Jack Preiss[2]

[1]Institute of Biological Chemistry
Washington State University
Pullman WA 99164-6340

[2]Biochemistry Department
Michigan State University
East Lansing, MI 48824

INTRODUCTION

For many crop plants, improvements in agricultural technology and the application of greater amounts of mineral nutrients no longer results in cost-effective enhancement of crop yield (King et al., 1967). In view of the public's concern on the release of chemicals in the environment and its likely impact on the use of chemical fertilizers and pesticides in agriculture, future increases in crop yields will be a result of genetic improvements of existing plants in photosynthesis and their ability to protect themselves against pest and pathogens. Since relatively minor increments in the efficiency of photosynthesis can result in dramatically increased crop yields, much of the current research has been focused toward directly altering the components of the photosynthetic apparatus and the CO_2 fixation enzyme activity, Ribulose bisphosphate carboxylase, and it ancillary oxygenase activity. Alternatively many scientists are examining a less direct approach of enhancing photosynthesis by increasing the capacity of plants to store photosynthate in its nutrient reserves.

Several studies (Ewart et al.; 1986, Nasyrov, 1978; Tollenaar, 1977) have shown that the rate and duration of photosynthesis in the leaf tissue (source) and the capacity of the storage tissue (sink) to assimilate photosynthate are interrelated. Under enviromental conditions where photosynthesis is not limiting, overall plant productivity appears to be regulated by the sink strength. For example, the removal of storage organs from wheat (King et al., 1967), cotton (McArthur et al., 1975), and potato (Burt, 1964) induces a general depression of photosynthetic activity in the leaves, whereas expansion of the sink tissue via root top grafts stimulates photosynthesis (Throne et al., 1964). Since starch is the major storage form of photosynthate in the sink tissue of many crop plants, enhancement in its biosynthesis may stimulate photosynthesis, and in turn, increase overall crop yields. Here we discuss the past and present evidence indicating a central role of ADPglucose pyrophosphorylase in regulating starch synthesis. From these we offer a rational basis for manipulating its genes as a means to enhance photosynthesis and improve overall plant productivity, and the biochemical and molecular properties of the enzyme and corresponding genes.

PATHWAY OF STARCH SYNTHESIS

An abundance of evidence indicates that the ADPglucose pathway consisting of the enzymatic reactions catalyzed consecutively by ADPglucose pyrophosphorylase (EC 2.7.7.27; reaction 1), starch synthase (EC 2.4.1.21; reaction 2) and branching enzyme (EC 2.4.1.18; reaction 3) is the predominant pathway if not the sole pathway towards starch synthesis .

(1) Glucose-1-P+ATP <——> ADPGlc+PPi

(2) ADPGlc + (α-1,4-glucosyl$)_n$ <——> ADP + (α-1,4-glucosyl$)_{n+1}$

(3) (α-1,4-glucosyl$)_{n+1}$ <——> (α-1,4; α1,6-glucosyl$)_n$

The data supporting this pathway are from a number of biochemical and genetic studies. As only a brief discussion will be presented here, the interested reader is referred to several recent reviews (Preiss, 1982; 1988; 1991; Preiss and Levi, 1979; 1980) for more detailed discussion. Studies of mutants from maize (Cameron et al 1954; Creech 1965; Tsai et al. 1966; Dickinson et al. 1969; Hannah et al., 1980), pea (Smith et al., 1989), and *Arabidopsis* (Lin et al., 1988a; 1988b) showed reduced amounts of ADPglucose pyrophosphorylase activity leads to reduced.starch levels. Mutations at the independent loci Shrunken-2 and Brittle-2 result in dramatic reduction of maize endosperm ADPglucose pyrophosphorylase activity as well as approximately 25% of normal starch content (Cameron et al., 1954; Creech, 1965; Tsai, 1966; Hannah et al., 1980). Recently, it has been shown that the *rb* locus controls the level of ADPglucose pyrophosphorylase in developing pea embryos (Smith et al., 1989). Pea lines homozygous for the recessive allele, *rb*, displayed a 28 to 62% reduction of starch in the pea embryo compared to the wildtype. Concommitantly, there was a 95 to 97% reduction in ADPglucose pyrophosphorylase activity without affect upon the activities of the other enzymes of starch biosynthesis. A mutant of *Arabidopsis thaliana* (Lin et al., 1988a,b), containing less than 0.2% of the leaf starch content observed in the normal strain had only 0.2% of the normal leaf ADPglucose pyrophosphorylase activity without alterations in the activities of the other starch biosynthetic enzymes.

Biochemical studies have also shown that the kinetic properties of the enzymes in the ADPGlucose pathway (K_m and V_{max} values) are consistent with the known concentrations of substrate and effector metabolites present in the various plant cells in contrast to the known properties of the UDPglucose specific starch synthase and plant phosphorylases (Preiss et al., 1982; 1988; 1991, Preiss and Levi, 1979; 1980). For these latter two enzymes the high K_m values for the respective substrates (UDPglucose and glucose-l-P) compared to the actual cellular levels preclude a significant role, if any, in starch synthesis. In view of the fact that phosphorylase catalyzes an equilibrium reaction in cells having intracellular Pi concentrations in excess of glucose-l-P it is most likely that phosphorylase plays a role in starch degradation rather than synthesis (Preiss et al., 1982; 1988; 1991, Preiss and Levi, 1979; 1980). Both biochemical and genetic evidence suggests that the ADPglucose pathway previously depicted is the major route for starch biosynthesis.

REGULATION OF ADPGLUCOSE PATHWAY

ADPglucose pyrophosphorylase from plants is allosterically regulated; the enzyme is activated by 3-phosphoglyceric acid (3-PGA) and inhibited by ortho-phosphate (Pi). Several lines of evidence support the allosteric regulation of ADPglucose pyrophosphorylase as the dominant control of starch biosynthesis. Many of these studies have evaluated the effect of Pi levels on the rate of starch biosynthesis. Santarius and Heber (1965) have shown that the concentration of Pi in leaf chloroplasts ranges from 5-10 mM and decreases by 30-50% in the light. Assuming that the 3-PGA concentration remains unchanged at 5 mM under these light-dark conditions, *in vitro* kinetic data indicates that the rate of ADPglucose synthesis would increase 5- to 23-fold in the light (Ghosh and Preiss, 1966; Copeland and Preiss, 1981; Morell et al., 1987). Since other glycolytic intermediates and ATP undoubtedly increase during photosynthesis, the rate of ADPglucose synthesis is probably much higher. Other studies (Mares et al., 1978; Steup et al., 1976; Kaiser et al., 1979a; 1979b) also support the role of Pi in regulating starch biosynthesis. Steup et al. (1976) have demonstrated that low concentrations of Pi have little effect on the rate of CO_2 fixation in illuminated spinach leaf chloroplasts, whereas higher concentrations (1mM) not only induced a long lag in starch biosynthesis, but also suppressed the maximum rate attained. Extensive studies by the laboratories of Heldt, Heber and Walker (Heldt et al., 1977) revealed that starch synthesis in illuminated chloroplasts was inhibited by Pi, a phenomenon which could be reversed by the addition of 3-PGA in the medium. Through measurement of metabolite concentrations under such experimental conditions, these workers were able to show that the rate of starch synthesis corresponded directly with the ratio of 3-PGA to Pi in the stroma and, more importantly, agreed very well with the *in vitro* activity of ADPglucose pyrophosphorylase at these different 3-PGA to Pi ratios (Ghosh and Preiss, 1966; Copeland and Preiss, 1981; Morell et al., 1987). The role of Pi in regulating starch biosynthesis via ADPglucose pyrophosphorylase can also be inferred from studies on starch accumulation in wheat leaves infected by *Puccinia striiformis* (MacDonald et al., 1970) and mannose feeding of leaf dics (Shen-Hwa et al., 1975; Herold et al., 1975). In both instances, the sequestration of Pi by the pathogen (for utilization for its own metabolic needs), or through the formation of an unmetabolizable end product (mannose l-P), the result was elevated levels of starch biosynthesis. The overwhelming evidence mentioned supports the hypothesis that the allosteric behavior of ADPglucose pyrophosphorylase responds to the fluctuating *in vivo* ratio of 3-PGA to Pi and plays a dominant role in the regulation of starch biosynthesis (Preiss, 1982; 1988; 1991; Preiss and Levi, 1979; 1980).

The availability of chloroplastic mutants of phosphoglucose isomerase of *Clarkia xantiana* (Kruckenberg et al., 1989; Neuhaus et al., 1988) and *Arabidopsis thaliana* mutants of phosphoglucomutase (Caspar et al.,1986) and ADPglucose pyrophosphorylase (Lin et al., 1988a; 1988b; Neuhaus et al., 1990), has shed light on the extent these enzymes exert on chloroplast starch synthesis. Studies conducted with the *Clarkia* mutants showed that reduction of plastid phosphoglucoisomerase had little effect on starch synthesis in low light, but in saturating light showed a 50% reduction in starch synthesis with no corresponding increase in sucrose synthesis. In *Clarkia* the reduced levels of the cytosolic phosphoglucoisomerase (18% of wildtype) lowered the sucrose synthetic rates and increased the rate of starch synthesis. Metabolite levels were also affected in these mutants. In the mutant containing only 18% wildtype cytosolic phosphoglucoisomerase activity, both fructose 2-6-P_2 and 3-PGA levels increased 100%. Neuhaus et al. (1989) concluded that their results provide strong evidence that reduced sucrose synthesis rate is due to the increased fructose 2-6-P_2 levels which inhibits the cytosolic fructose-1,6-bisphosphatase, a key regulatory enzyme in sucrose synthesis. They also indicate that the data strongly support the view that increased starch synthesis in the mutants with reduced levels of phosphoglucoisomerase is due to activation of the ADPglucose pyrophosphorylase by the increased 3-PGA concentrations. Thus these experiments provide furthur support for the *in vivo* regulation of starch synthesis by 3-PGA.

Neuhaus and Stitt (1990) utilized the null chloroplast phosphoglucomutase (Caspar et al., 1986) and the low activity (7% of wildtype levels) ADPglucose pyrophosphorylase mutants (Lin et al., 1988) of *Arabidopsis thaliana* to construct hybrid plants containing 50% of wildtype levels of phosphoglucomutase activity and 50% of wildtype levels of ADPglucose pyrophosphorylase activity, respectively. The effects of these reduced activities on CO_2 fixation, and starch and sucrose fluxes in low and high light intensity were measured. In low light, a 50% decrease in phosphoglucomutase activity had no significant effect on the above fluxes. A 50% and 93% decrease of ADPglucose pyrophosphorylase activity, however, resulted in a 23% and 74% decrease, respectively, in starch synthesis with a concomitant increase of 17% and 42% in the rate of sucrose synthesis. Thus, diminution of ADPglucose synthesis activity not only significantly affected starch synthesis but also affected the partitioning of photosynthetic carbon, causing more to be directed towards sucrose biosynthesis. In high light, a 50% decrease in phosphoglucomutase activity resulted in a 20% decrease in starch synthesis with little effect on the rate of sucrose synthesis. In contrast, reduction of the ADPglucose synthesizing activity by 50% and 93%, however, resulted in a 39% or 90% decrease in starch biosynthesis flux. The flux of photosynthetic carbon under these conditions was not redirected towards sucrose synthesis but rather the photosynthetic rate was reduced about 46%. The flux coefficients for the enzymes for starch biosynthesis were calculated to determine the distribution of control. Of the enzymes under study, the flux to starch synthesis is only regulated by ADPglucose pyrophoshorylase in low light. In high light and CO_2, although ADPglucose pyrophosphorylase activity exerts major control, other enzymes such as plastid phosphoglucomutase and phosphoglucoisomerase also exert control, although not to the same extent..

In contrast to the extensive studies on leaf ADPglucose pyrophosphorylases, less is know about the physical and kinetic properties of the enzyme from storage tissues (i.e., seeds and tubers). The potato tuber enzyme has allosteric specificities similar to the spinach leaf enzyme (Sowkinos 1976, 1981; Sowokinos and Preiss, 1982). Early observations of the maize endosperm enzyme (Dickinson et al., 1969b; Amir et al., 1972) showed that the ADPglucose pyrophosphorylase activity was allosterically-activated by 3-PGA and inhibited by Pi; however, their reponse to these effectors was not as large as that seen in leaf tissue (Ghosh et al, 1966). Maximal activation of these enzymes was observed at rather high concentrations of 3-PGA (2-10 mM), with only a 3- to 4-fold activation. Recent studies from the Preiss laboratory indicate that ADPglucose pyrophosphorylases from nonphotosynthetic tissues are less stable than their counterparts from leaf tissue and, when appropriate precautions are used, the enzymes are fully responsive to allosteric regulation *in vitro*. Plaxton et al. (1987) have shown that the endosperm-specific enzyme from maize tissue is highly sensitive to proteolysis, resulting in an enzyme less responsive to allosteric activation by 3-PGA. Inclusion of protease inhibitors during enzyme purification preserves the integrity of the enzyme. Using the intact enzyme an approximate 25-fold activation was detected with 5 mM 3-PGA, a response nearly identical to that detected with the enzyme from leaf tissue (Ghosh et al., 1966). Moreover, the maximum activity of the potato tuber enzyme is only observed in the presence of 3 mM dithiothreitol. In the absence of reducing agent, the enzyme shows a 4-fold reduction on activity (Sowokinos and Preiss, 1982). The enzyme from non-photosynthetic sink organs appears to be more sensitive to proteolysis and oxidation than the leaf enzyme.

Although ADPglucose pyrophoshorylase from leaf and nonphotosynthetic tissue show similar responses to allosteric activation and inhibition by small effector molecules, an important difference in catalytic properties between these enzymes is their turnover rate for ADPglucose synthesis. The purified spinach leaf enzyme has a specific activity of about 150 units/mg protein (Morell et al., 1987) to 200 units/mg protein (K. Ball and J. Preiss, unpublished observation), whereas the highly purified potato tuber enzyme is about 1/3 of this value (Okita et al., 1990). The lower specific activity of the tuber enzyme may be compensated for by the overall mass of this enzyme during tuberization. Indeed, the most important aspect of starch biosynthesis in storage tissues such as potato tubers (Sowokinos, 1981), maize seeds (Ozbun et al., 1973), and wheat seeds (Turner, 1969; Jenner, 1982; Reeves et al., 1986) may be in regulating the synthesis of the starch biosynthetic enzymes, such as ADPglucose pyrophosphorylase. In most of these studies, ADPglucose pyrophosphorylase activity or antigen level increased in correlation with the amount of starch or dry weight during the early stages of development.

STRATEGIES TO MANIPULATE ADPGLUCOSE PYROPHOSPHORYLASE FOR ENHANCED STARCH PRODUCTION

Starch is the major storage form of photosynthate in many agronomic field crops, e.g. potato, cereals etc., and can represent over 80% of the storage organ's dry weight (Jenner 1982). Although the physiological processes of loading, transport, and unloading of assimilate from the vascular system (Ho, 1988) can limit the rate of starch formation in nonphotosynthetic tissues, it is clear, as discussed above, that starch formation is centrally controlled by the catalytic activity of ADPglucose pyrophosphorylase (Preiss, 1982; 1988; Preiss and Levi, 1979; 1980). This is well established in leaf chloroplasts where during photosynthesis, 3-PGA levels increase due to CO2 fixation while Pi levels decrease because of corresponding increases in the levels of Calvin cycle intermediates as well as elevation in the levels of ATP due to photophosphorylation (Heldt et al., 1977). The increase in the 3PGA to Pi ratio results in conditions conducive toward an active form of ADPglucose pyrophosphorylase promoting the formation of ADPglucose, the substrate for starch synthase (Preiss, 1982; 1988; Preiss and Levi, 1979; 1980). In the dark, Pi levels increase with corresponding decreases in 3-PGA levels which result in an inactive form of the enzyme and thereby suppress ADPglucose formation. These changes in the stromal 3PGA/Pi ratios and their effect on ADPglucose pyrophosphorylase, therefore, readily account for the diurnal oscillation of starch synthesis observed in leaf tissue. In nonphotosynthetic storage tissue such as potato tubers and cereal seeds, the contribution by allosteric regulation of this enzyme in starch synthesis is not known. In contrast to the diurnal flucuations in 3-PGA and Pi in leaf chloroplasts, it is difficult to envision that the levels of these metabolites would change in the non-photosynthetic amyloplast. *In vitro* kinetic data of the potato tuber pyrophosphorylase, indicates that at equal concentrations of 3-PGA and Pi (levels typically seen in almost all systems so far studied), the rate of ADPglucose formation would be only 20% when compared to studies conducted in the absence of Pi (Sowokinos and Preiss, 1982). Some indirect evidence has been obtained that Pi may control ADPglucose pyrophosphorylase in non-photosynthetic tuber tissue. Hnilo *et al.*, (1989) observed a 50% increase in the incorporation of ^{14}C-labeled sucrose into starch in the presence 10 mM mannose, a Pi sequestrator. The results are consistent with the view that the reduction of intracellular Pi levels resulted in an increase in the catalytic activity of the endogenous ADPglucose pyrophosphorylase, which, in turn, increased the flow of carbon into starch. These results are consistent with the observed *in vitro* responses of the ADPglucose pyrophosphorylase (Sowokinos and Preiss, 1982) and suggests that maximal carbon flow into starch has not been achieved in the potato tuber and likely in the differentiated storage organs of other plants.

Since ADPglucose pyrophosphorylase plays such a pivotal role in starch biosynthesis and is catalytically regulated at multiple levels, it is a prime target for attempts to enhance the sink strength of storage organs. With respect to this discussion, sink strength could be potentially increased through (l) elevation of ADPglucose pyrophosphorylase gene expression (2) introduction of a pyrophosphorylase gene encoding a more catalyticallyactive enzyme and/or (3) introduction of a pyrophosphorylase gene that encodes an enzyme refractory to Pi.

Although direct evidence has yet to be obtained that an engineered ADPglucose pyrophosphorylase can enhanced starch production in the storage organs of higher plants, results from biochemical and genetic studies in the bacteria, *Escherichia coli,* supports this approach. In several instances, the cellular, molecular and biochemical aspects of bacterial glycogen biosynthesis is similar to the events that occur in the nonphotosynthetic storage tissue of higher plants. The biosynthesis of bacterial glycogen uses enzymes analogous to those found in higher plants, namely ADPglucose pyrophosphorylase, a 1,4-a-D-glucan: glucose 1-P adenylytransferase (glycogen synthase), and a 1,4-a-D-glucan: 1,4-a-D-glucan 6-a-D-(1,4-a-D-glucano)-transferase (branching enzyme). The *E. coli* ADPglucose pyrophosphorylase is also allosterically regulated by small effector molecules, although by fructose 1,6-bisphosphate and AMP. The differences in allosteric effectors between the plant and bacterial enzyme reflect the principal assimilatory pathway of carbon utilization found in these organisms, i.e. Calvin cycle (plants) vs. glycolysis (enteric bacteria). In addition to a common biosynthetic pathway, a-glucan synthesis in *E. coli* is tightly controlled during normal bacterial growth. The rate of glycogen biosynthesis is inversely related to the growth rate and increases dramatically during the negative log to stationary phase of growth in the presence of excess carbon (Fig. 1). During this period, the levels of ADPglucose pyrophosphorylase and glycogen synthase activities increase to their maximum levels due to the enhanced

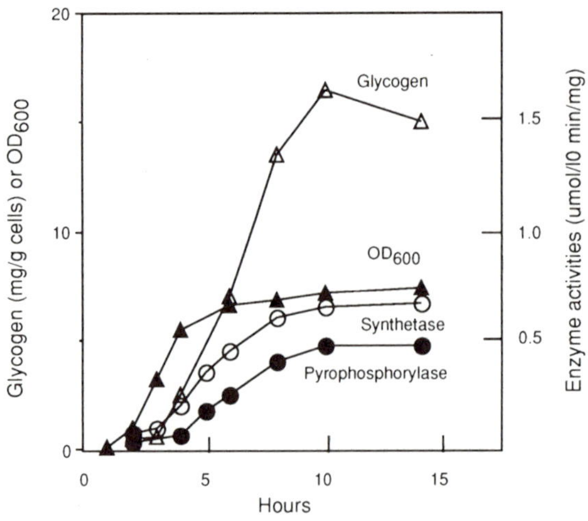

Fig. 1. Increase in glycogen, ADPglucose pyrophosphorylase and a-1,4-glucan synthetase (glycogen synthase) during growth of *Escherchia coli B*.

transcription of these genes. This control by transcription is acheived by the cyclic AMP-receptor complex and by guanosine S'-diphosphate 3'diphosphate, synthesized by the product of the *relA* gene (Preiss, 1984: Preiss and Romeo, 1989; Romeo and Preiss, 1989; 1990). Although comparable information on the molecular mechanisms governing control of the genes encoding the starch biosynthetic enzymes has not been obtained, the higher plant ADPglucose pyrophosphorylase genes are also controlled during plant growth (Jenner, 1982) and by excess carbon (Rober-Muller et al., 1990). During development of the potato tuber or cereal seeds, starch synthesis is readily evident in storage cells which differentiate after extensive cell division (Jenner, 1982). This is clearly evident in developing wheat endosperm where the rate of starch synthesis begins to accelerate in 6 day old of mRNA transcripts for ADPglucose

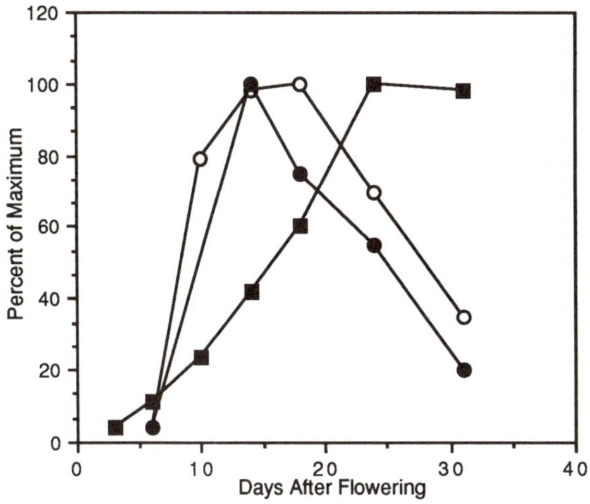

Fig. 2. Increase in ADPglucose pyrophosphorylase mRNA transcripts (l), antigen (o), and seed dry weight (n) during wheat seed development.

pyophosphorylase can be detected (Fig. 2). The rate of starch synthesis is highest between 10-14 days when the levels of ADPglucose pyrophosphorylase transcripts and protein are at their highest. In addition to these biochemical and molecular similarities, the starch containing amyloplasts, like *E. coli,* displays a heterotrophic pattern of growth dependent on both a carbon and energy source from the cytoplasm. Recent studies (Keeling et al., 1988) in wheat indicate that hexose monophosphates may be directly assimilated by the amyloplasts which could serve as a direct substrate for starch biosynthesis and energy source via glycolysis.

In view of the similarities between *E. coli* and higher plant amyloplasts, information obtained from analysis of the bacterial system may be directly applicable to higher plants. Extensive genetic and biochemical analysis by the Preiss laboratory (Preiss, 1978, 1984; Preiss and Romeo, 1990) has resulted in the isolation of a number of mutations that affect bacterial glycogen synthesis. Two classes of mutations that result in enhanced glycogen synthesis are of interest here. One class of glycogen excess mutations are those which affect the allosteric regulation of ADPglucose pyrophosphorylase. Three mutants SG5, CL1136, and 618 accumulate about three times more glycogen than the wildtype *E. coli B*. Enzyme kinetic analysis of the ADPglucose pyrophosphorylase from these mutants show that the concentration of the activator fructose 1,6-bisP required for 50% maximal activation ($A_{0.5}$) is about 3-fold less for the SG5 enzyme, about 4.5-fold for the mutant 618 enzyme, and less than 12-fold for the CL1136 enzyme, than that needed by the wildtype enzyme (Table 1). The mutant ADPglucose pyrophosphorylases are also less sensitive to the allosteric inhibitor AMP as higher concentrations are required to give 50% inhibition compared to the wildtype enzyme. The greater glycogen accumulation in these mutants is due to a alteration in the enzyme that causes a greater affinity for the activator and a lower affinity for the inhibitor. A second class of excess glycogen mutations are the genetic regulatory type as exemplified by the mutant strains SG3 and AC70Rl which contain defects at the regulatory loci *glgR* and *glgQ*, respectively (Preiss and Romeo, 1990). Mutations in *glgR* results in a 2-3 fold

Table 1. Comparison of glycogen accumulation and kinetic constants for allosteric effectors of ADPglucose pyrophosphorylases of *Escherichia coli* mutants with the parent wild-type strain (from Preiss and Romeo, 1990).

Organism	Glycogen accumulation (mg g^{-1})[a]	Fructose 1,6-bisphosphate, $A_{0.5}$ (μM)[b]	AMP $I_{0.5}$ (μM)[b]
Escherichia coli	20	68	75
Mutant CLl 136	74	5.2	680
Mutant SG5	35	22	170
Mutant 618	70	15	860

a The data shown above was obtained by growth of the organisms in minimal media with 0.75% glucose as the carbon source and is expressed as mg of anhydroglucose per gram (wet wt) of cells and is the maximal amount seen in stationary phase.

b $A_{0.5}$ is the concentration giving 50% of maximal stimulation aid $I_{0.5}$ is the concentration of AMP required for 50% inhibition in the presence of 1.5 mM fructose 1,6-bisphosphate for the *Salmonella typhimurium* enzymes.

elevation in glycogen biosynthesis and in the specific activities of ADPglucose pyrophosphorylase and glycogen synthase but not of branching enzyme. Mutations in *glgQ* results in even higher amounts of glycogen production and 6 to 10 times higher levels of the glycogen biosynthetic enzymes. As the kinetic constants for the substrates, activators and inhibitors of the ADPglucose pyrophosphorylase enzyme observed in these mutants were similar to the wildtype enzyme, the synthesis of excess glycogen was due solely to the overproduction of the glycogen biosynthetic enzymes, especially for ADPglucose pyrophosphorylase. These examples in a-glucan biosynthesis in bacteria support the premise that genetic manipulation of either the structural or regulatory genes of ADPglucose pyrophosphorylase may be a viable approach in alterating starch levels in higher plants. These molecular approaches, however, require an understanding of the structure of the ADPglucose pyrophosphorylase genes, how these genes are controlled during development, and the structure-function relationship of amino acid residues involved in enzyme catalysis and allosteric regulation.

ENZYME STRUCTURE

Higher Plants

Extensive genetic and biochemical studies from the Preiss laboratory have demonstrated that the bacterial ADPglucose pyrophosphorylase from both E. *coli* and *Salmonella typhimurium is* encoded by a single gene locus, *glgC* (Preiss, 1984). This gene encodes an polypeptide of 48 kD, which aggregates to form a homotetramer of about 200 kD. The higher plant enzyme was thought to have a structure similar to the bacterial enzyme, as molecular weight of the holoenzyme was first estimated to be about 200 kD (Ghosh et al., 1966). A subsequent study by Copeland and Preiss (1981), who purified the spinach leaf enzyme to apparent homogeneity, showed that under denaturing conditions, two subunits were present at equimolar amounts. It was not clear at that time whether the two subunits represented distinct polypeptides, or whether the smaller species was simply a degradation product of the larger subunit. More recent studies have now demonstrated that these subunits are non-identical, and are products of two different genes. Morell et al. (1987) were able to purify both the 51 and 54 kD subunits and showed that these subunits could be distinguished on the basis of their N-terminal amino acid sequences, peptide mapping patterns, and cross-reactivity to different polyclonal antibodies raised against each individual subunit. Moreover, mutants deficient in ADPglucose pyrophosphorylase have been obtained in *Arabidopsis* (Lin et al., 1988a; 1988b). In one mutant, *adg2,* leaf extracts contained only about 5% of the wildtype levels of ADPglucose pyrophosphorylase and were deficient in the 54 kD subunit when analyzed by

Western blot analysis (Lin et al., 1988b). A second mutation, *adg1*, affects the levels of both subunits, and genetic analysis suggests that this locus may be involved in regulation of gene expression (Lin et al., 1988a). Biochemical and genetic evidence indicates that the leaf specific ADPglucose pyrophosphorylase from the leaves of higher plants is composed of two distinct subunits, presumably encoded by two separate genes.

The presence of two distinct subunits raises the question of whether both polypeptides are necessary for enzymatic activity or whether two unique populations of enzyme exist, each composed solely of one type of subunit. In a recent study, Li and Preiss (1991) have purified the ADPglucose pyrophosphorylase from TL46, an *Arabidopsis* mutant which contained only 5% enzyme activity but accumulated 46% of starch as the wildtype. Like the wildtype enzyme, the mutant ADPglucose pyrophosphorylase activity displayed a native molecular weight of about 200,000 and was activated by 3-PGA and inhibited by Pi. The mutant enzyme, however, required higher concentrations of 3-PGA for activation and was more sensitive to inhibition by Pi than the wildtype enzyme. The mutant enzyme also displayed significantly less affinity toward the substrates, Glc 1-P and ATP as well as Mg^{2+} than the normal enzyme. Examination of the tertiary and quaternary structure showed only one immunoreactive band at 48 kD, in contrast to the wildtype enzyme which exhibits a 54 kD and 48 kD immunoreactive polypeptides. On native polyacrylamide gels, two immunoreactive bands were observed but only the faster migrating species exhibited enzyme activity. The slower immunoreactive band presumably was a precursor to the faster moving species or-contained improperly assembled small subunits. These results show that the small subunit was capable of assembling into an enzymatically active homotetrameric complex, albeit with altered enzyme kinetics. Moreover, as the mutant plants contained only 5% of the wildtype levels of ADPglucose pyrophosphorylase activity, it appeared that the presence of both subunits are required for full enzyme activity as exhibited by wildtype plants.

Previously, the structure of the ADPglucose pyrophosphorylase from nonphotosynthetic tissue was believed to be composed of a single type subunit analogous to the condition found in bacteria cells. Evidence for this was the results obtained by Sowokinos and Preiss (1982) who showed that the purified ADPglucose pyrophosphorylase from potato tuber appeared to be composed of a single subunit of about 50 kD as viewed by single dimensional SDS polyacrylamide gel electrophoresis. This conclusion was supported by the work of Krishnan et al. (1986) in which a comparison of crude extracts from endosperm tissue of rice, wheat, and maize by Western blot analysis using the antibody to the spinach leaf holoenzyme showed a single band, whereas examination of leaf tissue from these plants and spinach showed two dissimilar sized polypeptides. These results were consistent with the view that the leaf specific ADPglucose pyrophosphorylase was composed of two distinct subunits while non-photosynthetic tissue was a homotetrameric species. More recent studies has now indicated that the higher plant enzyme, irrespective of whether it is from leaf or non-photosynthetic tissue, is composed of two non-identical subunits. Okita et al. (1990) re-investigated the structure of the potato tuber enzyme. The enzyme was extensively purified using a modified protocol described by Sowokinos and Preiss (1982) coupled with Mono-Q FPLC chromotography (Okita et al.,1990). This enzyme preparation, which was estimated to be approximately 80% pure, was analyzed by SDS-PAGE to reveal a single dominant band of approximately 50 kD in agreement with the earlier study (Sowokinos and Preiss (1982). The pattern of enzyme activity following elution from the Mono-Q ion exchange column was rather broad suggesting that the subunits of this enzyme were heterogeneous in charge. This was confirmed by high resolution 2-D PAGE which resolved the purified tuber ADPglucose pyrophosphorylase into two polypeptide groups present in equimolar amounts based on the intensity of Coomassie blue staining. These two groups, besides having different net charges, also had slightly different molecular weights of approximately 50 and 51 kD. Western blot analysis of the 2-D gel, using antisera prepared against the purified tuber holoenzyme showed cross reactivity with both subunits. Antiserum raised against the spinach 51 kD small subunit cross-reacted strongly with the tuber small subunit (50 kD). Antiserum raised against the spinach leaf large subunit did not bind significantly to either potato tuber subunit. These observations suggested that the potato tuber enzyme consists of two immunologically distinct subunits.

Extensive genetic and biochemical evidence has now been obtained that endospermspecific ADPglucose pyrophosphorylase is composed of two distinct subunits and products of unique genes (Preiss et al., 1990; Bae et al., 1990; Bhave et al., 1990). Extensive purification of the maize endosperm enzyme showed three prominent bands of 98 kD, 60 kD, and 55 kD on SDS polyacrylamide gels with only the 55 kD polypeptide cross-reacting with the antibody raised against the spinach leaf holoenzyme (Plaxton et al., 1987), an observation in agreement with the earlier results of Krishnan et al. (1986). The identification of the other ADPglucose pyrophosphorylase subunit was subsequently obtained by Western blot analysis using affinity-purified antibodies raised against the large and small spinach leaf enzyme (Preiss et al., 1990). Using subunit-specific antibodies, it was readily demonstrated that the antibody raised against the spinach leaf 51 kD subunit cross-reacted only with the maize 55 kD polypeptide, while the anti-spinach leaf 54 kD subunit cross-reacted with the maize 60 kD subunit. The reactivity of the spinach leaf anti-54 kD subunit to the maize 60 kD polypeptide was much weaker than reaction displayed by the spinach anti-S 1 kD subunit to maize 55 kD polypeptide, suggesting that the large subunits of the spinach and maize enzyme were more divergent than the small subunits. This assignment of the 60 kD and 55 kD subunits as being the constituents of the maize enzyme was supported by the analysis of genetic mutants. In maize, two mutations, *Brittle2* (Bt_2) and *Shrunken-2* (Sh_2) have been isolated which are defective in ADPglucose pyrophosphorylase activity and starch production (Creech, 1965; Tsai et al., 1966; Dickinson et al., 1969; Hannah et al., 1975; Hannah et al., 1980; Tuschall et al., 1982). Analysis of these mutants revealed that Bt_2 was missing the 55 kD species while the Sh_2 mutant lacked the 60 kD polypeptide (Preiss et al., 1990). The identification of the maize ADPglucose pyrophosphorylase subunits was substantiated by the structural analysis of recombinant cDNA clones. Barton et al (1986) obtained cDNAs, pES6-66 and pES6-75, which encoded the 60 kD and 55 kD polypeptides, respectively. Significant hybridization of pES6-75 was demonstrated to a rice endosperm cDNA (Preiss et al. 1990) known to encode the small subunit sequences (Anderson et al, 1989). Moreover, the derived primary sequences of these two plasmids and those subsequently isolated revealed substantial homology to known ADPglucose pyrophosphorylase sequences from both plants and bacteria (Bhave et al., 1990; Bae et al., 1990). Based on these results, pES6-75 (Bt_2) and pES6-66 (Sh_2) contain the structural gene sequences for the maize endosperm ADPglucose pyrophosphorylase small and large subunit, repectively. Overall, the results of these studies indicate that the higher plant ADPglucose pyrophosphorylase, irrespective of its tissue origin, is composed of two non-identical subunits encoded by distinct genes.

Cyanobacteria

The cyanobacteria are procaryotic organisms which perform oxygenic photosynthesis and CO_2 fixation similar to the chloroplasts of higher plants. This semblance is also reflected in oc-glucan synthesis whose properties are a hybrid between enteric bacteria and higher plants. As shown in many bacteria, these organisms accumulate glycogen but their ADPglucose pyrophosphorylase is activated by 3-PGA and inhibited by Pi (Levi and Preiss, 1976; Iglesias et al. 1991). The allosteric regulation by these small effector molecules is consistent with the utilization of the Calvin pathway of CO_2 fixation observed in these organisms (Preiss, 1978). In view of the evolutionary relationship of these organisms to both bacteria and higher plants, Iglesias et al. (1991) studied the regulatory and structural properties of the enzymes from the filamentous, heterocystic, *Anabaena* PCC 7120 and unicellular *Synechocystis* PCC 6803. Both enzymes were strongly activated by 3-PGA and inhibited by Pi suggesting that the ratio of 3-PGA to Pi could play a prominent role in the regulation of glycogen synthesis in these organisms. Pyruvate and AMP did not stimulate or inactivate, respectively, either enzyme while both enzymes were only marginally activated by fructose 1,6-bisP or fructose 6-P. The absence or reduced responses to these metabolites indicate that the regulatory properties of the cyanobacterial ADPglucose pyrophosphorylase differed substantially from those displayed by the enteric bacterial enzymes.

The purified enzymes from *Synechocystis* and *Anabaena* displayed a single polypeptide of 53 kD and 50 kD, respectively, on both single- and two-dimensional polyacrylamide denaturing gels. The molecular weight of the native enzymes, as determined by gel filtration chromatography, were 216,000 *(Synechocystis)* and 225,000 *(Anabaena)*, indicating a homotetrameric structure for both cyanobacteria enzymes. The homotetrameric structure of the cyanobacteria enzymes was confirmed by the isolation of the structural gene and its expression in E. coli to yield active enzyme activity (Charng, Kakefuda and Preiss, unpublished data). Thus the cyanobacteria enzymes display a homotetrameric structure common to other bacterial enzymes but nevertheless possesses regulatory properties of those from the higher plants.

COMPARISON OF PLANT AND BACTERIAL ADP GLUCOSE PYROPHOSPHORYLASE SEQUENCES

To gain insight into the structure/function relationship of the plant enzyme, cDNA clones which encode the two subunits responsible for the heterotetrameric structure of the potato tuber enzyme have been obtained (Anderson et al., 1990; Nakata et al., 1991). A comparison of the deduced amino acid sequence of the small and large potato subunits to the *E. coli* subunit (Nakata et al., 1991; du Jardin and Berhin, 1991) displayed an overall identity of 37% and 35% respectively. Similar observations have been made with the primary sequences of the large and small subunit of the maize endosperm enzyme (Bae et al., 1990; Bhave et al., 1990). Despite this low overall homology, the domains which are important for allosteric and catalytic activity of the bacterial enzyme are more conserved (Smith-White and Preiss, 1991). Lys_{39} and Lys_{195} of the *E. coli* sequence have been shown to be protected from phosphopyridoxylation by the activator fructose-1,6-bisP and substrate ADPglucose suggesting these two residues are located within or near the allosteric and substrate binding sites, respectively (Baecker et al., 1983; Parsons and Preiss, 1978). These two regions are highly conserved in the potato small subunit and less so in the large subunit (Fig. 3). This is readily evident by the replacement of the conserved Lys_{110} in the potato small subunit sequence by an arginine in the large species. The role of Lys_{195} has been determined for the *E. coli* enzyme by site-specific mutagenesis (Hill et al., 1991). Replacement of the this residue of Lys_{195} of the *E. coli* enzyme produced enzyme activities whose Km values for glucose 1-P ranged from 100- to 10,000-fold greater than the wildtype enzyme depending on the nature of residue replacement. Kinetic constants measured for ATP, Mg^{2+}, and the allosteric activator, however, remained unchanged as compared to the wildtype enzyme. These results indicate a role for Lys_{195} in the binding of glucose 1-P (Hill et al., 1991). Although the role of the aligned Lys_{39} has not been identified in the plant enzyme, the significant degree of homology in this region suggests it may contribute to the stability of the protein conformation and possibly to the regulatory and/or catalytic functions of the plant enzyme. Also of interest is Tyr_{114} of the *E.coli* enzyme which has been shown to be involved in substrate binding (Lee et al., 1986). This region is highly conserved in both the small and large subunits of the potato enzyme (SmithWhite and Preiss, 1991); however, in the plant subunits the Tyr is replaced by Phe. To test whether the phenolic hydroxyl group played a critical role in enzyme catalysis, Kumar et al.(1988) replaced Tyr_{114} with Phe in the bacterial enzyme via site directed mutagenesis. This mutation caused a lowering of the apparent affinity of the enzyme for substrate but does not appear to be obligatory for catalysis . The altered enzyme also displayed decreased affinity toward its activator. This conservation of a hydrophobic aromatic amino acid at this position may be of importance in coordinating the adenine rings of ATP and ADPglucose via hydrophobic interactions. Photoaffinity studies have been conducted with the spinach enzyme using 8-N_3ADPglucose (Preiss et al., 1989). Incorporation is seen mainly if not solely in the large subunit. It will be interesting to see if it labels the area homologous to the bacterial substrate binding region.

Since the bacterial and plant enzymes catalyze the same chemical reaction their structural dissimilarities most likely reflect differences in their specificity toward the allosteric effectors (Anderson et al., 1989; Smith-White and Preiss, 1991). This is readily apparent in the peptide region containing the AMP inhibitor binding site of the E. *coli* enzyme (Larsen et al., 1986; Smith-White and Preiss, 1991). A direct comparision of the plant and bacterial sequences show significant differences (Fig. 4). Some alignment of conserved residues is evident but a 13 residue insertion/deletion must be introduced in the upstream sequence and two insertions/deletions of the same resulting size in the downstream flanking sequence to accommodate the alignment. The result is movement of the conserved domain relative to the enzyme's active site. As the plant enzyme is insensitive to AMP, it is likely that the structural differences evident in this region is at least partially responsible for the change in allosteric specificity between the bacterial and plant enzymes. Moreover, Morell et al. (1988) have shown that pyridoxal phosphate which mimics the effector molecule 3-PGA equally labels both subunits of the spinach enzyme. For the small subunit, the pyridoxal phosphate binding region has been determined and the reactive Lys residue identified (Morell et al., 1988). This Lys residue and surrounding region is situated near the C-terminus of the plant enzyme and is identically conserved in the small subunit of all the known plant enzymes (Nakata et al., 1991; Smith-White and Preiss, 1991). This region is also highly conserved in the large subunit. Recent results (Ball and Preiss, unpublished observations), however, indicate that

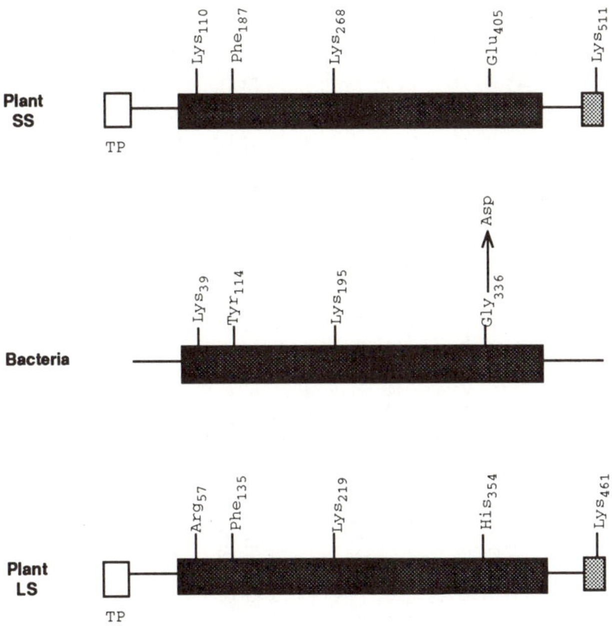

Fig. 3. Schematic representation of the primary structures of the plant and bacterial ADPglucose pyrophosphorylase subunits. The two conserved regions are shown as black boxes. Residues essential for allosteric control or substrate binding of the bacterial enzyme and the corresponding plant residues are illustrated. The open boxes represent the transit peptides required for plastid targeting of the plant subunits. The regions containing Lys_{511} and Lys_{461} of the plant large (LS) and small (SS) subunits are conserved regions thought to be involved in the allosteric regulation of the plant enzymes. Lee et al. (1987) and Kumar et al. (1988) have shown when an Asp is substituted for Gly_{336} the mutant enzyme exhibits altered allosteric properties. Numbering of the plant LS is based on the known derived primary sequence obtained from a truncated cDNA clone.

```
              235        245       255       265       275          285                 295
E. coli        SK                 SLASMGIYVFDADYLYELLEEDDRDENSSHDFGKDLIPKITEAGLAYAHPFPL SCVQSDPDAEP
SL-54kD       DTTVLGLSK....                          ...FPTA*DFGSEII   *A...                ...AYLFND
WL7           .........EE  AEKKPYI****V*I*KKEI*LN**RWRFPTA*DFGSEII   *AAAREIN             **KAYLFND
Sh-2          ETNFLSYAIDD  AQKYPYL********KK*A*LD**KSKYTQLHDFGSEIL   *RAVLDHS              **ACIFTG
WE7           DTSFLNFAIDD  PAKYPYI****V***KR*V*LN**KSRYAELHDFGSEIL   *RALHDHN              **AYVFTD
WE3           DTSFLNFAIDD  PAKYPYI********KR*V*LN**KSRYAELHDFGSEIL   *RALHDHN              **AYVFTD
pot-51 kD      TTLVGLSPQD  AKKSPYI****VY**KT*V*LK**KWSYPTSNDFGSEII   *AAIDDYN              **AYIFKD
rice seed     DTTILGL   DDVRAKEMPYI*******ISKNVMLQ**REQFPGA*DFGSEVI   *GA*NI*M            R**AYLYDG
pot-50 kD     DTTILGL   DDKRAKEMPFI*******ISK*VMLN**RDKFPGA*DFGSEVI   *GA*SL*M            R**AYLYDG
B 22-1        DTTILGL   DDKRAKEMPFI*******ISK*VMLN**RDKFPGA*DFGSEVI   *GA*SL*M            R**AYLYDG
SL-51 kD      DTTILGL   DDERAKEMPYI*******ISK*VMLN**RDKFPGA*DFGSEVI   *GA*SL*L            R**AYLYDG
A. thaliana   DTTILGL   DDERAKEMPFI*******VSR*VMLD**RNQFPGA*DFGSEVI   *GA*PL*L            R**AYLYHG
bt-2          DTTILGL   DDVRAKEMPYI********SK*VMLQ**REQFPEA*DFGSEVI   *GA*SI*K            R**AYLYHG
```

Figure 4. Comparison of the primary structures of all of the known ADPGlc pyrophosphorylases relative to that of the peptide of the *E. coli* protein covalently modified by azido-AMP, a small molecule probe of the allosteric inhibitor binding site of the *E. coli* enzyme. The integers atop the group correspond to the number of residues of the rice seed small subunit sequence. The residues of the other ADPGlc pyrophosphorylase proteins which are identical to that found in the *E. coli* are indicated as a star. The peptide identified as being covalently modified in the *E. coli* protein is emboldened. Large subunit sequences are for spinach leaf 54 kD (SL-54kDa), wheat leaf (WL7), maize *Shrunken-2* (sh-2), and wheat endosperm (WE7 and WE3) and potato (pot 51 kD).

three Lys residues in the spinach leaf large 54 kD subunit may be involved in the binding of 3-PGA. Pyridoxal-P, the 3-PGA analogue, binds to two Lys residues close to the C-terminus and also to a Lys residue near the Nterminal region. The sequences are, S-G-I-T-V-I-F-K*-Q-A-T-I-K-D-G-V-V, I-K-D-A-I-I-D-K*-N-A-R, and K*-W-F-Q-G-T-A-D-A-V-R. The astericks indicating the labeled lysine residues. The relative locations of these peptides are based on similar conserved amino acid sequences in the small and large subunits of the ADPglucose pyrophosphorylases from rice seed, spinach leaf, potato tuber, maize endosperm, and wheat endosperm (Preiss, 1991; Smith-White and Preiss, 1992). Of interest is that the presence of 1 mM 3-PGA prevents incorporation of [^{3}H]-pyridoxal-P into all Lys residues in the large and small subunits (Ball and Preiss, unpublished observations). Pi, however, inhibits only incorporation into the small subunit and only one of the Lys residues of the large subunit (I-K-D-A-I-I-D-K*-N-A-R). Thus all the Lys residues may be involved in 3-PGA binding but with respect to Pi, its binding is at or near the labeled Lys residue of the small subunit and only one of the labeled Lys residues of the large subunit. Overall, these observations suggests that the allosteric region of the plant enzyme contains and essential peptide sequence located mainly in its C-terminus rather than in its N-terminus as in the bacterial enzyme. The difference in allosteric specificity is probably the result of mutations in other regions of the plant enzyme. Lee et al. (1987) and Kumar et al. (1988) have shown when Gly_{336} of the bacterial enzyme is changed to the acidic residue, Asp, the bacterial enzyme becomes less responsive towards allosteric activation by fructose-1,6-P_2. It is interesting to note that the amino acid at the comparable position in the potato, rice, and spinach small subunits is negatively charged, but not in the large subunits of the potato (Nakata et al., 1991), and wheat endosperm and leaf enzymes (Olive et al., 1989) which display a His. An acidic, Asp, is present, however, in *Shrunken-2* which encodes the large subunit of the maize endosperm enzyme. Based on the evidence presently available, it appears that the change in allosteric specificity between the bacterial and plants enzymes is due to specific segmental and point mutations at several different locations in the coding sequences of these genes.

EVOLUTION OF THE ADPGLUCOSE PYROPHOSPHORYLASE GENES

A comparison of the small and large potato tuber subunits revealed a 52% amino acid identity overall (Nakata et al., 1991) which is somewhat higher than the 35% identity observed between the spinach leaf subunits (Preiss et al., 1991). The potato small subunit showed 84% identity to the rice endosperm small subunit (Anderson et al., 1990) and 93% identity to the spinach leaf small subunit (Preiss et al., 1991) on the primary level. The potato large subunit exhibited 58% and 61% amino acid identity to the wheat endosperm AGA.7 (Olive et al., 1989) and maize endosperm shrunken-2 (Bhave et al., 1990) clones respectively. There appears to be less identity between the two subunits from the same species than between similar subunits from the different plants (Nakata et al., 1991). The homology studies also suggest that the smaller subunits are more conserved among the different plants and the larger subunits more divergent.

These preliminary observations have now been confirmed by the study of Smith-White and Preiss (1991) who made detailed comparisons of the DNA and primary sequences between the enteric bacteria and the available plant enzyme sequences. In addition to showing that many domains are retained in the proteins from both bacteria and higher plants, the proteins from angiosperms can be readily classified into two main groups encoding the small and large subunits with the large subunit group further subdivided into two subgroups. Of the two groups, the one comprised of the small subunit genes have diverged only slightly over time. Substantial sequence identity is evident between the rice endosperm, maize endosperm (Bt_2), potato tuber, spinach leaf, and *Arabidopsis* enzyme. A weaker similarity is observed between the sequences for the maize endosperm large subunit (Sh_2), potato tuber large subunit, and the putative large subunit of the wheat endosperm enzyme, and between the wheat leaf and spinach leaf large subunits. Although the degree of homology among components within these two subgroups of the large subunit genes is significantly less than that evident among members of the conserved small subunit genes, a large number of positions in the primary structure of these proteins show conserved residues which are unique to these subgroups.

The subdivision of the large subunit genes into two groups appears to correlate with the tissue type (photosynthetic versus nonphotosynthetic) from which the mRNA was originally isolated, suggesting that there are two general types of enzymes, leaf-specific and storage organ-specific (Smith-White and Preiss, 1991). Each of the enzyme types displays the conserved small subunit while the large subunit would have a more plastic structure which may reflect some specific role in metabolism. Immunological analysis of the ADPglucose pyrophosphorylase enzyme from different tissues of the same plant support this conclusion of Smith-White and Preiss (1991). Antibody raised against the small subunit of the spinach leaf enzyme strongly cross reacts with the small subunit of the maize endosperm (Preiss et al 1990), rice endosperm (Anderson et al., 1990; Lee et al., 1987), and wheat endosperm (Okita et al., 1990) enzymes, as well as to the leaf-specific small subunit from these plants. In contrast, antibody specific to the spinach leaf large subunit reacts weakly, if at all, to the large subunits from the endosperm-specific enzyme, whereas significant reactivity to this antibody is displayed by the leaf counterparts.

Based on the low but certain homology between the large and small subunit primary sequences, these genes probably arose from a duplication of a common ancestor gene with subsequent divergent. Recent evidence indicating that the cyanobacterial enzyme, which has higher plant allosteric specificity, is a homotetramer like the bacterial enzyme (Iglesias et al., 1991) supports this concept. Moreover, the significant cross-reactivity of the cyanobacterial enzyme to the antibody specific to the spinach leaf small subunit (Iglesias et al., 1991) suggests that the conserved small subunit, rather than the large subunit, is closer to the archetypic gene. In view of the fact that the large subunit sequences between the photosynthetic leaf and non-photosynthetic storage organ have diverged to the same extent from the conserved small subunit sequences, Smith-White and Preiss (1991) have suggested that these two subgroups of the large subunit diverged shortly after the divergence of the large from the small subunit.

DEVELOPMENTAL REGULATION AND GENE ISOLATION

The developmental expression of the gene encoding the small subunit of rice (Anderson et al., 1991), potato (Anderson et al., 1990), and the analogous gene from wheat (Reeves et al., 1986) has been examined. Results from preliminary experiments indicate that the pattern of mRNA accumulation closely follows the level of ADPglucose pyrophosphorylase enzyme activity and starch accumulation. This observation suggests that this gene is controlled transcriptionally. Northern blot and Western blot data (Anderson et al., 1990) suggests that the same potato small subunit gene is expressed in both leaf and tuber. In contrast, Krishnan et al. (1986) and Olive et al. (1989) have shown in rice and wheat that there are different genes encoding the same type subunit which are expressed either in leaf or endosperm tissue.

Recent studies by Muller-Rober et al. (1990) have shown that the regulation of the large and small subunit genes from potato is complex as the patterns of expression in different tissues and under different environmental conditions differ significantly. In flowers, leaves of different developmental stages, and middle stem sections, there were significantly more transcripts for the small subunit than for the large subunit. In contrast, equivalent levels of transcripts for both subunits were detected in lower stem and root tissue, while the large subunit mRNAs showed a higher steady state level than the small subunit transcripts in developing tubers. This difference in developmental patterns of their transcripts was also reflected in their capacity to be induced by sucrose. mRNAs for the large subunit showed a dramatic increase when leaf petioles were incubated in sucrose, whereas those for the small subunit remained largely unchanged. The expression of the ADPglucose pyrophosphorylase genes, particularly the one that encodes the large subunit, has a major impact on the capacity of tissues to synthesize starch.

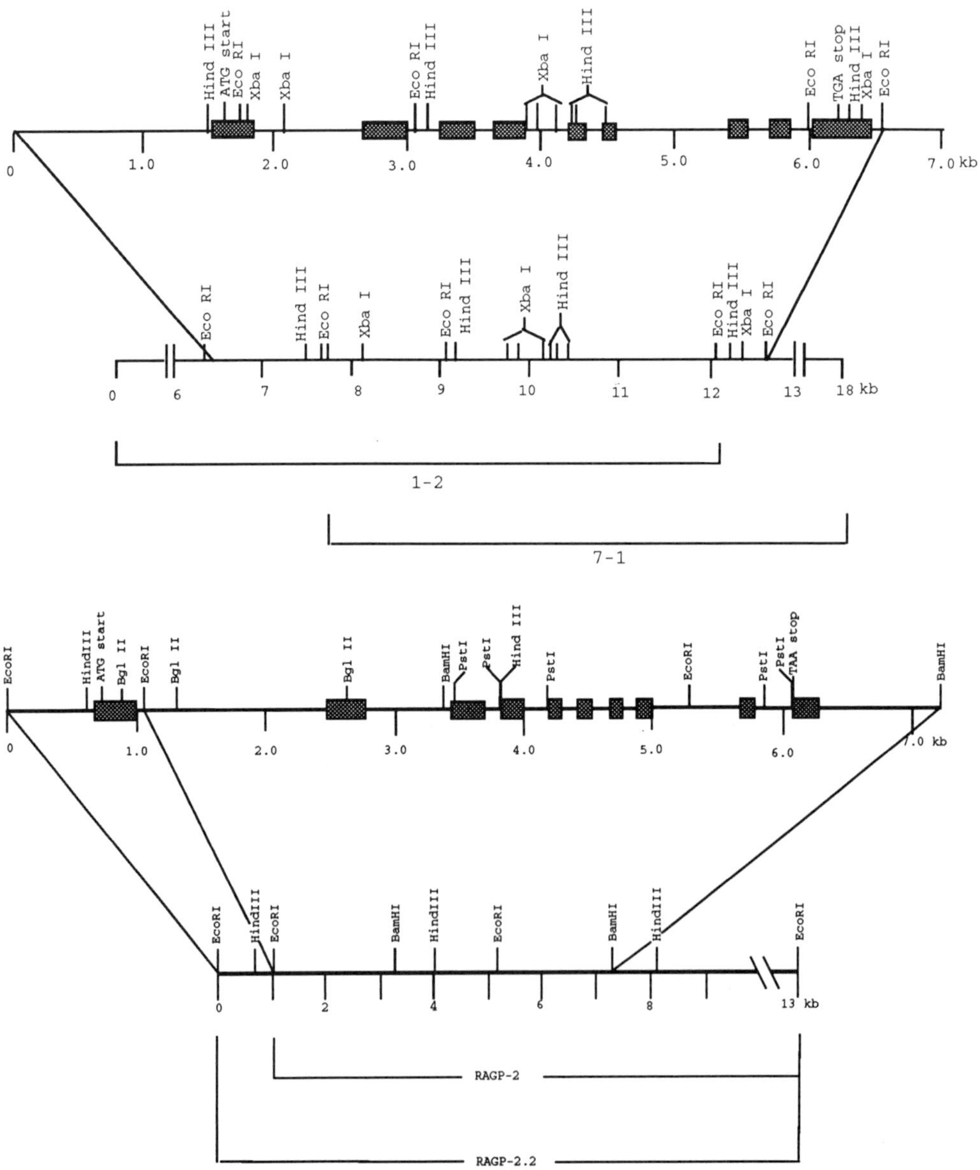

Fig. 5. Physical maps of the restriction sites of the potato tuber and rice endosperm small subunit genes and their spatial relationship to the coding region. Exons are represented by the black boxes. Both maps were derived from two overlapping genomic clones as indicated.

Southern blot hybridization analysis shows that the gene encoding the tuber pyrophosphorylase small subunit is contained within the nucleus in 1-2 copies/haploid genome (Anderson et al., 1990). The structure of the genes encoding the rice endosperm and potato tuber small subunits have been recently determined (Anderson et al., 1991; Nakata et al., in press) and spans approximately 6.5 and 5.5 kb respectively. Unlike most plant genes, the small subunit ADPglucose pyrophosphorylase genes from rice and potato tuber are highly complex in structure with the coding sequences interrupted by 9 and 8 introns respectively (Fig. 5). The introns vary in size between the two genes, but interrupt the respective coding region at the same locations. The only exception, intron 9 of the rice gene, is inserted 3 amino acids from the stop codon in a region not contained in the potato gene. The complexity of the ADPglucose pyrophosphorylase gene structure is also shared by two other genes that encode enzymes involved in starch metabolism. These genes, one for sucrose synthase (Werr et al., 1985) and the other for the granule bound starch synthase (Klosgen et al., 1986), displayed a relatively complex gene structure possessing 16 and 14 exons, respectively. The presence of multiple introns contained within the ADPglucose pyrophosphorylase gene as well as other genes involved in starch metabolism suggests some role in the regulation of gene expression as observed for the first intron of the alcohol dehydrogenase gene (Callis et al.,1987) and sucrose synthase gene (Maas et al., 1991).

SUMMARY

The study of photosynthesis and its enhancement is of extreme interest in the development of more efficient crop plants. Increasing the sink strength (the capacity to assimilate or store photosynthate) is a feasible way of increasing the photosynthetic capacity of plants that utilize starch as the major storage form of photosynthate. The overwhelming evidence indicates that the central enzyme controlling the sink capacity in such plants is ADPglucose pyrophosphorylase. A notable feature of this enzyme is its regulatory properties. The activity of this enzyme from both leaf and most nonphotosynthetic tissues is stimulated by 3-PGA and inhibited by Pi. During diurnal growth, the ratio of these small effector molecules fluctuate in leaf chloroplasts, which modulate the catalytic activity of ADPglucose pyrophosphorylase and hence affect the rate of carbon flow into starch. Regulation of starch synthesis by ADPglucose pyrophosphorylase activity is also believed to occur in nonphotosynthetic tissue such as the potato tuber and cereal seeds, although to what extent allosteric regulation plays a role in modulating starch synthesis has yet to be directly evaluated. Based on genetic analysis of a-glucan synthesis in *E. coli*, the overproduction of ADPglucose pyrophosphorylase and/or an enzyme with altered allosteric characteristics may improve the conversion of photosynthate into starch in terminal sink tissue. Such a condition would potentially relieve any possible feedback effects by small metabolites on photosynthesis which, in turn, would result in an increase in overall plant productivity.

ACKNOWLEDGEMENTS

Supported by grants from the National Science Foundation (DMB 86-10319) and USDA/DoE/NSF Plant Science Center Program (88-37271-3964) to J.P., the Department of Energy (DE-FG06-87ER13699) and Project 0590, College of Agriculture and Home Economics, Washington State University, Pullman, WA 99164-6340 to T.W.O, and a McKnight Foundation Fellowship to P.A.N.

REFERENCES

Amir, J., and Cherry, J.H., 1972, Purification and properties of adenosine diphosphoglucose pyrophosphorylase from sweet corn, *Plant Physiol.* 49:893-897.

Anderson, J.M., Okita, T.W. and Preiss, J., 1990, Enhancing carbon flow into starch: the role of ADPglucose pyrophosphorylase, *in:* "The Molecular and Cellular Biology of the Potato," M.E. Vayda, and W.D. Park, eds., C A B International, Wallingford, UK.

Bae, J.M., Giroux, J., and Hannah, L., 1990, Cloning and characterization of the *Brittle-2* gene of maize, Maydica 35:317-322.

Baecker, P.A., Furlong, C.E., and Priess, J., 1983, Biosynthesis of bacterial glycogen; primary structure of *Escherichia coli* ADPglucose synlhetase as deduced fromnucleotide sequence of the *glgC* gene, J. *Biol. Chem.* 258:5084-5088.

Barton, C., Yang, L., Galvin, M., Sengupta-Gopalan, C., and Borelli, T., 1986, Isolation of the shrunken2 and brittle-2 genes from maize. *in:* "Regulation of Carbon and Nitrogen Reduction and Utilization in Maize," J.C. Shannon, D.P. Knievel and C.D. Boyer, eds., American Society of Plant Physiology, Rockville, MD, pp. 363-365.

Bhave, M.R., Lawrence, S., Barton, C., and Hannah, C., 1990, Identification and molecular characterization of *shrunken-2* cDNA clones of maize, *Plant Cell* 2:581-588.

Briarty, L.G., Hughes, C.E., and Nevers, A.D., 1979, The developing endosperm of wheat: a stereological analysis, *Annals Bot.* 44:641-658. Burt, R.L., 1964, Carbohydrate utilization as a factor in plant growth, *Aust.* J. *Biol. Sci.* 17:867-877.

Callis, J., Fromm, M., and Walbot, V., 1987, Introns increase gene expression in cultured maize cells, *GenesandDevelop.* 1:1183-1200. Cameron, H.W., and Teas, H.J., 1954, Carbohydrate relationships in developing and mature endosperms of brittle and related maize genotypes, *Am.* J. *Bot.* 41:50-55.

Caspar, T., Huber, S.C., and Somerville, C., 1986, Alterations in growth, photosynthesis, and respiration in a starchless mutant of *Arabidopsis thaliana* (L.) deficient in chloroplast phosphoglucomutase activity, *Plant Physiol.* 79:11-17.

Copeland, L., and Preiss, J., 1981, Purification of spinach leaf ADPglucose pyrophosphorylase, *Plant Physiol.* 68:996-1001. Creech, R.G., 1965, Genetic control of carbohydrate synthesis in maize endosperm, *Genetics* 52:11751 186.

Dickinson, D.B., and Preiss, J., 1969, Presence of ADPglucose pyrophosphorylase in *shrunken-2* and *brittle-2* mutants of maize endosperm, *Plant Physiol.* 44:1058-1062.

Dickinson, D.B., and Preiss, J., 1969b, ADPglucose pyrophosphorylase from maize endosperm, *Arch. Biochem. Biophys.* 130:119- 128.

du Jardin, P., and Berhin, A., 1991, Isolation and sequence analysis of a cDNA clone encoding a subunit of the ADPglucose pyrophosphorylase of potato tuber amyloplasts, *Plant Mol. Biol.* 16:349-351.

Ewart, A.J., 1986, On assimilatory inhibition in plants, *J. Linnean Soc. Bot.* 31:364-461.

Ghosh, H.P., and Preiss, J., 1966, Adenosine diphosphate glucose pyrophosphorylase. A regulatory enzyme in the biosynthesis of starch in spinach chloroplasts, *J. Biol. Chem.* 241:4491-4505.

Hannah, L.C., and Nelson, O.E., 1975, Characterization of adenosine diphosphate glucose pyrophosphorylases from developing maize seeds, *Plant Physiol.* 55:297-302.

Hannah, L.C., Tuschall, D.M., and Mans, R.J., 1980, Multiple forms of maize endosperm ADPglucose pyrophosphorylase and their control by *shrunken-2* and *brittle-2, Genetics* 95:961-970.

Heldt, H.W., Chon, C.J., Maronde, D., Herold, A., Stankovic, Z.S., Walker, D.A., Kraminer, A., Kirk, M.R., and Heber, U., 1977, Role of orthophosphate and other factors in the regulation of starch formation in leaves and isolated chloroplasts, *Plant Physiol.* 59:1146-1155.

Herold, A., Lewis, D.H., and WaLker, D.A., 1975, Sequestration of cytoplasmic orthophosphate by mannose and its differential effect on photosynthetic starch synthesis in C3 and C4 species, *New Phytol.* 76:397-407.

Hnilo, J., Okita, T.W., 1989, Mannose feeding and its effect on starch synthesis in developing potato tuber discs, *Plant and Cell Physiol.* 30:1007-1010.

Hill, M.A., Kaufmann, K., Otero, J., and Preiss, J., 1991, Biosynthesis of bacterial glycogen: mutagenesis of a catalytic site residue of ADP-glucose pyrophosphorylase from *Escherchia coli. J. Biol. Chem.* 266: 12455-12460.

Ho, L.C., 1988, Metabolism and compartmentalization of imported sugars in sink organs in relation to sink strength, *Ann. Rev. Plant. Physiol. Plant Mol. Biol.* 39:355-378.

Iglesias, A.A., Kakefuda, G., and Preiss, J., 1991, Regulatory and structural properties of the cyanobacterial ADPglucose pyrophosphorylases, Plant Physiol. in press.

Jenner, C.F.,1982, Storage of starch, *in:* "Encyclopedia of Plant Physiology: Plant Carbohydrates I," Vol 13A, F.A. Loewus, and W. Tanner, eds., Springer-Verlag, Berlin, pp.700-747.

Kaiser, W.M., and Bassham, J.A., 1979, Light-dark regulation of starch metabolism in chloroplasts I. Level of metabolites in chloroplasts and medium during light-dark transition *Plant Physiol.* 63:105-108.

Kaiser, W.M., and Bassham, J.A., 1979, Light-dark regulation of starch metabolism in chloroplasts. II. Effect of chloroplastic metabolite levels on the formation of ADPglucose by chloroplast extracts, *PlantPhysiol.* 63:109-113.

Keeling, P.L., Wood, J.R., Tyson, R.H., and Bridges, I.G. 1988, Starch synthesis in developing wheat grain: evidence against the direct involvement of triose phosphates in the metabolic pathway. Plant Physiol. 87: 311-319.

King, R.W., Wardlaw, I.F., and Evans, L.T., 1967, Effect of assimilate utilization on photosynthetic rate in wheat, *Planta* 77:261-276.

Klosgen, R.B., Gierl, A., Schwartz-Sommer, Z., and Saedler, H.,1986, Molecular analysis of the *waxy* locus of *Zea mays, Mol. Gen. Genet.* 203:237-244.

Krishnan, H.B., Reeves, C.D., and Okita, T.W., 1986, ADPglucose pyrophosphorylase is encoded by different mRNA transcripts in leaf and endosperm of cereals, *Plant Physiol.* 81:642-645.

Kruckeberg, A.L., Neuhaus, H.E., Feil, R., Gottlieb, L.D., and Stitt, M., 1989, Decreased-activity mutants of phosphoglucose isomerase in the cytosol and chloroplast of *Clarkia xantiana.* Impact on mass-action ratios and fluxes to sucrose and starch and estimation of flux control coefficients andelasticitycoefficients, *Biochem. J.* 261:457-467.

Kumar, A., Tanaka, T., Lee, Y.M., and Preiss, J., 1988, Biosynthesis of bacterial glycogen: use of site-directed mutagenesis to probe the role of tyrosine^{114} in the catalytic mechanism of ADPglucose synthetase from *Escherichia coli, J. Biol. Chem.* 263:14634-14639.

Larsen, C.E., Lee, Y.M., Preiss, J., 1986, Covalent modification of the inhibitor-binding site(s) of *Escherchia coli* ADP-glucose synthetase, *J. Biol. Chem.* 261:15402-15409.

Lee, Y.M., Kumar, A., and Preiss, J., 1987, Amino acid sequence of *Escherichia coli* ADPglucose synthetase allosteric mutant as deduced from the DNA sequence of the *glgC* gene, *Nucleic Acids Res.* 15:10603.

Li, L., and Preiss, J., 1991, Characterization of ADPglucose pyrophosphorylase from a starch deficient mutant of *Arabidopsis thaliana* (L.), Carbohydrate Res. in press.

Lin, T.P., Caspar, T., Somerville, C., and Preiss, J., 1988a, Isolation and characterization of a starchless mutant of *Arabidopsis thaliana* (L.) Henyh lacking ADPglucose pyrophosphorylase activity, *Plant Physiol.* 86:1131-1135.

Lin, T.P., Caspar, T., Somerville, C., and Preiss, J., 1988b, A starch deficient mutant of *Arabidopsis thaliana* with low ADPglucose pyrophosphorylase activity lacks one of the two subunits of the enzyme, *PlantPhysiol.* 88:1175-1181.

Maas, C., Laufs, J., Grant, S., Korfhage, C., Werr, W., 1991, The combination of a novel stimulatory element in the first exon of the maize *shrunken-1* gene with the following intron 1 enhances reporter gene expression up to 1000-fold, *Plant Mol. Biol.* 16:199-207.

MacDonald, P.W., and Strobel, G.A., 1970, Adenosine diphosphate glucose pyrophosphorylase control of starch accumulation in rust-infested wheat leaves, *Plant Physiol.* 46:126-135.

Mares, D.J., Hawker, J.S., and Possingham, J.V., 1978, Starch synthesizing enzymes in chloroplasts of developing leaves of spinach *(Spinacea oleracea* L.), *J. Exp. Bot.* 29:829-835.

McArthur, J.A., Hesketh, J.D., and Baker, D.N., 1975, Cotton, *in:* "Crop physiology: some case histories, " L.T. Evans, ed., Cambridge University Press, London.

Morell, M., Bloom, M., Knowles, V., and Preiss, J., 1987, Subunit structure of spinach leaf ADPglucose pyrophosphorylase, *Plant Physiol.* 85: 185-187.

Morell, M., Bloom, M., and Preiss, J., 1988, Affinity labeling of the allosteric activator site(s) of spinach leaf ADPglucose pyrophosphorylase, *J. Biol. Chem.* 263:633-637.

Nakata, P.A., Greene, T.W., Anderson, J.M., Smith-White, B.J., Okita, T.W., and Preiss, J., 1991, Comparison of the primary sequences of two potato tuber ADPglucose pyrophosphorylase subunits, *Plant Mol. Biol.* 17:1089-1093.

Nasyrov, Y.S., 1978, Genetic control of photosynthesis and improving of crop productivity, *Ann. Rev. Plant Physiol.* 29:215-237.

Neuhaus, H.E., Kruckeberg, A.L., Feil, R., and Stitt, M., 1989, Reduced-activity mutants of phosphoglucose isomerase in the cytosol and chloroplast of *Clarkia xantiana* II. Study of the mechanisms which regulate photosynthate partitioning, *Planta* 178:110-112.

Neuhaus, H.E., and Stitt, M., 1990, Control analysis of photosynthate partitioning: impact of reduced activity of ADPglucose pyrophosphorylase or plastid phosphoglucomutase on the fluxes to starch and sucrose in *Arabidopsis, Planta* 182:445454.

Okita, T.W., Nakata, P.A., Anderson, J.M., Sowokinos, J., Morell, M., and Preiss, J., 1990, The subunit structure of potato tuber ADPglucose pyrophosphorylase, *Plant Physiol.* 93:785-790.

Olive, M.R., Ellis, R.J., and Schuch, W.W. 1989, Isolation and nucleotide sequences of cDNA clones encoding ADPglucose pyrophosphorylase polypeptides from wheat leaf and endosperm, *Plant Mol. Biol.* 12:525-538.

Ozbun, J.L., Hawker, J.S., Greenberg, E., Lammel, C., Preiss, J., and Lee, E.Y.C., 1973, Starch synthetase, phosphorylase, ADPglucose pyrophosphorylase, and UDPglucose pyrophosphorylase in developing maize kernels, *Plant Physiol.* 51:1-5.

Parsons, T.F., and Preiss, J., 1978, Biosynthesis of glycogen: isolation and charactertization of the pyridoxal-P allosteric site and the ADPglucose-protected pyridoxal-P binding site of Escherchia coli B ADPglucose synthase. J. Biol. Chem. 253:7673-7645

Plaxton, W.C., and Preiss, J., 1987, Purification and properties of nonproteolytic degraded ADPglucose pyrophosphorylase from maize endosperm, *Plant Physiol.* 83:105-112.

Preiss, J.,1978, Regulation of ADPglucose pyrophosphorylase, *in:* "Advances in Enzymology and Related areas of Molecular Biology," Vol. 46, A. Meister, ed., John Wiley and Sons Inc., New York, pp. 317-381.

Preiss, J., 1984, Bacterial glycogen synthesis and its regulation, *Ann. Rev. Microbiol.* 38:419-458.

Preiss, J., 1982, Regulation of the biosynthesis and degradation of starch, *Ann. Rev. Plant Physiol.* 33:419-458.

Preiss, J., 1988, Biosynthesis of starch and its regulation, *in:* "The Biochemistry of Plants," Vol. 14, J. Preiss, ed., Academic Press, New York, pp. 181-254.

Preiss, J., Biology and molecular biology of starch synthesis and its regulation, *in:* "Oxford Survey of Plant Molecular and Cellular Biology," Vol. 7, B.J. Miflin, ed., Oxford University Press, Oxford, In Press.

Preiss, J., Ball, K., Hutney, J., Smith-White, B., Li, L., and Okita, T.W., 1991, Regulatory mechanisms involved in the biosynthesis of starch, *Pure Appl. Chem.* 63:535-544.

Preiss, J., Cress, D., Hutny, J., Morell, M., Bloom,M., Okita, T., and Anderson, J.,1989, Regulation of starch synthesis. Biochemical and genetic studies, *in:* "ACS Symposium Series 389 on Biocatalysis in Agricultural Biotechnoloty," J.R. Whitaker, and P.E. Sonnet, eds., American Chemical Society, Washington D.C., pp. 84-92.

Preiss, J., Danner, S., Summers, P.S., Morell, M., Barton, C.R., Yang, L., and Nieder, M., 1990, Molecular characterization of the *brittle-2* gene effect on maize endosperm ADPglucose pyrophosphorylase subunits, *Plant Physiol.* 92:881-885.

Preiss, J., and Levi, C., 1979, Metabolism of Primary Products of Photosynthesis: Metabolism of Starch in Leaves, *in* "Encyclopedia of Plant Physiology, New Series", Vol. 6, M. Gibbs and E. Latzko, eds., Springer-Verlag, New York, pp. 282-312.

Preiss, J., and Levi, C., 1980, Starch biosynthesis and degradation, *in:* "The Biochemistry of Plants," Vol. 3, J. Preiss, ed., Academic Press Inc, New York, pp. 371-423.

Preiss, J., and Romeo, T., 1989, Physiology, biochemistry and genetics of bacterial glycogen synthesis, in: "Advances in Microbial Physiology," Vol. 30, A. H. Rose, and D.W. Tempest, eds., Academic Press Limited, San Diego, CA, pp. 183-238.

Reeves, C.D., Krishnan, H.B., and Okita, T.W., 1986, Gene expression in developing wheat endosperm. Accumulation of gliadin and ADPglucose pyrophosphorylase messenger RNAs and polypeptides, *Plant Physiol.* 82:34-40.

Muller-Rober, B.T., Kobmann, J., Hannah, L.C., Willmitzer, L., Sonnewald, U., 1990, One of two different ADP-glucose pyrophosphorylase genes from potato responds strongly to elevated levels of sucrose, *Mol. Gen. Genet.* 224:136-146.

Romeo, T., and Preiss, J., 1989, Genetic regulation of glycogen biosynthesis in *Escherchia coli:* in vitro effects of cyclic AMP and guanosine 5'-diphosphate 3'-diphosphate and analysis of *in vivo* transcripts, *J. Bacteriol.* 171:2773-2782.

Romeo, T., and Preiss, J., 1990, Genetic regulation of glycogen biosynthesis in *Escherchia coli:* in vivo effects of the catabolite repression and stringent response systems in glg gene expression, *Current Microbiol.* 21 :1 31- 1 37.

Santarius, K.A., and Heber, U., 1965, Changes in the intracellular levels of ATP, ADP, AMP, and Pi and regulatory function of the adenylate system in leaf cells during photosynthesis, *Biochim. Biophys. Acta* 102:39-54.

Sanwal, G.G., Greenberg, E.,Hardie, J., Cameron, E.C. and Preiss, J., 1968, Regulation of starch biosynthesis in plant leaves: activation and inhibition of ADPglucose pyrophosphorylase *Plant Physiol.*, 43:417-427.

Shen-Hwa, C.S., Lewis, D.H., and Walker, D.A. 1975, Stimulation of photosynthetic starch formation by sequestration of cytoplasmic orthophosphate, *New Phytol.* 74:383-392.

Smith, A.M., Bettey, M., and Bedford, I.D., 1989, Evidence that the *rb* locus alters the starch content of developing pea embryos through an effect on ADPglucose pyrophosphorylase, *Plant Physiol.* 89:1279-1289.

Smith-White, B.J., Preiss, J., 1992, Comparison of proteins of ADP-glucose pyrophosphorylase from diverse sources, J. Mol. Evol. *in press*.

Sowokinos J.R., 1976, Pyrophosphorylases in *Solanum tuberosum* . I. Changes in ADPglucose and UDPglucose pyrophosphorylase activities associated with starch biosynthesis during tuberization, maturation, and storage of potatoes, *Plant Physiol.* 57:63-68.

Sowokinos J.R., 1981, Pyrophosphorylases in *Solanum tuberosum.* II. Catalytic properties and regulation of ADPglucose and UDPglucose pyrophosphorylase activities in potatoes, *Plant Physiol.* 68:924929.

Sowokinos J.R., and Preiss, J., 1982, Pyrophosphorylases in *Solanum tuberosum* III. Purification, physical and catalytic properties of ADPglucose pyrophosphorylase in potatoes, *Plant Physiol.* 69: 1459-1466.

Steup, M., Peavey, D.G., and Gibbs, M., 1976, The regulation of starch metabolism by inorganic phosphate, *Biochem. Biophys. Res. Commun.* 72:1554-1561.

Thorne, G.N., and Evans, A.F., 1964, Influence of tops and roots on net assimilation rate of sugar beet and spinach beet grafts between them, *Ann. Bot.* 28:499-508.

Tieman, D.M., and Handa, A.K., 1989, Immunocytologicalization of polygalacturonase in ripening tomato fruit, *Plant Physiol.* 90: 17-20.

Tollenaar, M., 1977, Sink-source relationships during reproductive development in maize: a review, Maydica 22:49-75.

Tsai, C.Y., and Nelson, O.E., 1966, Starch-deficient maize mutant lacking adenosine diphosphate glucose pyrophosphorylase activity, *Science* 151:341-343.

Turner, J.F., 1969, Starch synthesis and changes in uridine diphosphate glucose pyrophosphorylase and adenosine diphosphate glucose pyrophosphorylase in the developing wheat grain, *Aust. J. Biol. Sci.* 22:1321-1327.

Tuschall, D.M., and Hannah, L.C., 1982, Altered maize endosperm ADPglucose pyrophosphorylases from revertants of a *shrunken-2* dissociation allele, *Genetics* 100:105-111.

Werr, W., Frommer, W.B., Maas, C., and Starlinger, P., 1985, Structure of the sucrose synthase gene on chromosome 9 of *Zea mays* L., *EMBO J.* 4:1373-80.

GENE INTROGRESSION FOR COTTON IMPROVEMENT: CONTRAST OF TRADITIONAL BREEDING WITH BIOTECHNOLOGIES

David W. Altman

International Service for the Acquisition of Agri-Biotech Applications
1509 22nd Street
Washington, D.C. 20037-1073

INTRODUCTION

The New World tetraploid cottons have a long history of cultivation, and consequently the process of gene introgression has an ancient origin. The earliest date for archaeological remains of Upland cotton *(Gossypium hirsutum)*, the most widely cultivated species accounting for over 90% of world production (Lee, 1984), ranges from 3500 to 2300 B.C. from the Tehuacan Valley in Mexico (Smith and Stephens, 1971). Another cultivated tetraploid, *G. barbadense* with 8% of world production (Lee, 1984) and also known as Egyptian or Pima cotton, dates from 2500 to 1750 B.C. in the Ancon-Chillon district of Peru (Stephens and Moseley, 1974). The distributions of these two cotton species were known to overlap from pre-Columbian times (Lee, 1984). The result of this proximal development of two cultivated species was that gene introgression between these two cottons occurred frequently. Percy and Wendel (1990), from a recent isozyme study, found that 22% of improved cultivars of *G. barbadense* had introgressant alleles from *G. hirsutum,* and many wild accessions from a wide geographic range also had detectable introgression.

With this long history of natural introgression, it would be presumptuous to claim that biotechnology has made a large contribution to cotton gene introgression. What biotechnology does allow is the potential to introgress any characterized gene without consideration for usual biological barriers. Prior to modern plant breeding technology, cotton gene introgression was accomplished by happenstance when cross-pollination took place as described above. Cotton breeders and geneticists expanded that range to include other strains, races, and species within *Gossypium.* New biotechnologies, including tissue culture methods, first enlarged the

Gene Conservation and Exploitation
Edited by J.P. Gustafson *et al.*, Plenum Press, New York, 1993

possibilities within the genus. Now, researchers who utilize biotechnologies have eliminated normal biological barriers by creating transgenic plants. For various reasons, cotton probably will be one of the first crop plants to be commercialized as a result of this process.

This paper will contrast traditional gene introgression with gene introgression via biotechnology, using the example of cotton. I will not attempt a thorough historical review of cotton gene introgression, because several other references have covered aspects of this subject (Hutchinson, 1959; Fryxell, 1976, 1979; Narayanan et al., 1984; Meredith, 1991). This presentation will update the status of germplasm resources and discuss some examples of traditional breeding using these resources. Cell culture-assisted introgression will receive more in-depth review, including past contributions, and some recent data from our research group will be presented.

GERMPLASM RESOURCES

From my perspective, many ambiguities exist about the status of available cotton genetic resources. Wild populations of *Gossypium* are found throughout areas of the tropics and subtropics on five continents and Oceania. Some regions are still remote and/or not readily accessible. Research dollars, and consequently well-trained botanists, have been relatively scant for cotton systematics studies and germplasm maintenance. Alternatively, few botanists have chosen to concentrate efforts on *Gossypium,* possibly due to the complexity of taxonomic problems. The one tetraploid and at least six diploid genomes have incompletely characterized interrelationships, and new materials indicate a necessary expansion of previous delineations. Consequently, recent findings have not been integrated with previous information, and inaccurate or incomplete systematic treatments in earlier publications are perpetuated. With that preface, the reader should note that my compilation, with assistance from Paul Fryxell, of a list of the currently accepted species of *Gossypium* (Table 1) only represents present knowledge. Future revisions undoubtedly will be necessary. The current list includes 52 taxa, whereas the most recent systematic monograph (Fryxell, 1979) listed 39 taxa. The past decade has seen renewed interest, and subsequently more plant explorations, for discovering and identifying wild populations. The bulk of the newly discovered species are from areas in the Horn of Africa and in northwestern Australia.

Several countries maintain cotton germplasm in some fashion, most notably the U.S.A., the U.S.S.R., China, India, and France (including cooperative assistance with several African nations). Only one smaller collection officially contains materials from interspecific hybrids (Maréchal,

1983). The working collection of the U.S. National Collection of *Gossypium* Germplasm, with approximately 5,300 accessions, is housed in the USDA-ARS unit at College Station, TX under the direction of A.E. Percival. Most of these accessions have been catalogued with the available accompanying data (Percival, 1987), and the Collection includes *Gossypium* wild species, race collections of *G. hirsutum*, obsolete varieties of *G. hirsutum*, an Asiatic collection of *G. herbaceum* and *G. arboreum* varieties and lines, and a *G. barbadense* collection of wild strains, varieties, and marker stocks. Percival and Kohel (1990) reviewed the status of the Collection, some of the more recent plant explorations, and related topics.

From a practical standpoint, the U.S. Collection, as well as other national collections, do not have adequate seedstocks of all materials to allow the required access by breeders and geneticists for any type of introgression program. For example, at least 15 of the 52 taxa are not presently available for distribution from the U.S. Collection (Percival, pers. commun.), due to the lack of seed multiplication capabilities outside of native habitats, the absence of seed collection from plant exploration trips, seed inviability, or various other reasons. The obsolete variety collection is incomplete, depends entirely on donations from private companies or individuals, and can have limitations required by plant variety protection. Several important materials, including cytogenetic stocks and genetic marker stocks, are not officially part of the Collection, and the availability of these lines depends on public programs with virtually no direct maintenance funding. In addition to these restrictions, some of this germplasm requires special techniques to obtain adequate germination and/or reasonable plant growth and flowering, and seed availability might not assure plant availability for hybridization. Race conversion programs are ongoing to develop day-neutral strains of the wild races of *G. hirsutum* and *G. barbadense* at the USDA-ARS laboratories at Mississippi State, MS and Maricopa, AZ, respectively, as a method to provide greater accessibility to this portion of the U.S. Collection (McCarty et al., 1979; Percy and Turcotte, 1989).

Biotechnological advances are not extensive for cotton germplasm conservation and maintenance. Germplasm traditionally has been stored as seed, and field collecting was limited almost exclusively to seed. A simple tissue culture method has been developed for supplementing field collecting (Altman et al., 1990a), but this technique has not yet been utilized by many cotton collectors. Other micropropagation methods were also developed, such as shoot tip culture (Renfroe and Smith, 1986), which are difficult to perform outside of specialized laboratories. Some preliminary work has been published on pollen utilization with microspore culture (Barrow, 1986), but the established collections have not planned to develop pollen storage as a component of the germplasm conservation effort.

TRADITIONAL GENE INTROGRESSION

Except possibly for remote areas of the Florida Keys, Upland cotton did not have any endemic populations in the U.S.A. Certain Indian tribes (Hopi, Papago, and Pima) of the Southwest introduced cotton cultivation to the southwestern U.S.A. (Lewton, 1912; Jones, 1936), but these cottons were unknown to the early eastern American colonists. Therefore, unlike many of our other major crops, cotton was a plant that had to undergo a significant change to allow its use in monoculture. *Gossypium hirsutum* is a tropical, perennial shrub which has been transformed to adapt to annual cultivation and massive fruiting in a temperate region.

Upland cotton production was established in the southern American colonies by the beginning of the 18th century (Watt, 1907; Fryxell, 1968).

Table 1. Currently accepted species of *Gossypium*.

Gossypium Linnaeus Species	Original Distribution	Genomic Designation	Reference
Subgenus *Sturtia* (R. Brown) Todaro			
Section *Sturtia*:			
G. robinsonii F. Mueller	Australia	C_2	Fryxell, 1979
G. sturtianum J.H. Willis	Australia	$C_{1\text{-}s}$	Fryxell, 1979
G. nandewarense Derera	Australia	$C_{1\text{-}n}$	Fryxell, 1979†
Section *Grandicalyx* Fryxell:			
G. cunninghamii Tod.	Australia	--‡	Fryxell, 1979
G. costulatum Tod.	Australia	–	Fryxell, 1979
**G. nobile* Fryx., Craven & J.M. Stewart	Australia	–	Fryxell et al., 1992
G. pulchellum (C.A. Gardner) Fryx.	Australia	–	Fryxell, 1979
**G. enthyle* Fryx., Craven & J.M. Stewart	Australia	–	Fryxell et al., 1992
**G. marchantii* Fryx., Craven & J.M. Stewart	Australia	–	Fryxell et al., 1992
G. populifolium (Bentbam) F. Muell. ex Tod.	Australia	–	Fryxell, 1979
**G. londonderriensis* Fryx., Craven & J.M. Stewart	Australia	–	Fryxell et al., 1992
**G. rotundifolium* Fryx., Craven & J.M. Stewart	Australia	–	Fryxell et al., 1992
**G. exiguum* Fryx., Craven & J.M. Stewart	Australia	–	Fryxell et al., 1992
G. pilosum Fryx.	Australia	–	Fryxell, 1979
Section *Hibiscoidea* Tod.:			
G. australe F. Muell.	Australia	C_3	Fryxell, 1979
C. nelsonii Fryx.	Australia	–	Fryxell, 1979
G. bickii Prokhanov	Australia	G_1§	Fryxell, 1979

Subgenus *Houzingenia* (Fryx.) Fryx.			
Section *Houzingenia*			
Subsection *Houzingenia*:			
G. trilobum (DC.) Skovsted	Mexico	D_8	Fryxell, 1988
G. thurberi Tod.	Mexico, Arizona	D_1	Fryxell, 1988
Subsection *Integrifolia* (Tod.) Tod.:			
G. davidsonii Kellogg	Mexico	$D_{3\text{-}d}$	Fryxell, 1988
G. klotzschianum Andersson	Galapagos Islands	$D_{3\text{-}k}$	Fryxell, 1979
Subsection *Caducibracteolata* Mauer:			
G. harknessii Brandegee	Baja California	$D_{2\text{-}2}$	Fryxell, 1988
G. armourianum Kearney	Baja California	$D_{2\text{-}1}$	Fryxell, 1988
G. tumeri Fryx.	Sonora	–	Fryxell, 1988
Section *Erioxylum* (Rose & Standley) Prokh.			
Subsection *Selera* (Ulbrich) Fryx.:			
G. gossypioides (Ulbr.) Standl.	Oaxaca	D_6	Fryxell, 1988
Subsection *Erioxylum* (Rose & Standl.) Fryx.:			
G. aridum (Rose & Standl. ex Rose) Skov.	Mexico	D_4	Fryxell, 1988
**G. schwendimanii* Fryx. & S. Koch	Mexico	–	Fryxell, 1988
G. lobatum H. Gentry	Michuoacan	D_7	Fryxell, 1988
**G. laxum* L. Phillips	Guerrero	D_9	Fryxell, 1988
Subsection *Austroamericana* Fryx.:			
G. raimondii Ulbr.	Peru	D_5	Fryxell, 1979
Subgenus *Gossypium*			
Section *Gossypium*			
Subsection *Gossypium*:			
G. arboreum L.	--¶	A_2	Fryxell, 1979
G. herbaceum L.	--¶	A_1	Fryxell, 1979
G. herbaceum subsp. *africanum* (Watt) Vollesen	Africa	A_1	Vollesen, 1987
Subsection *Anomala* Tod.:			
G. anomalum Wawra ex Wawra & Peyritsch	Africa	B_1	Fryxell, 1979
**G. senarense* Fenzl ex Wawr. & Peyr.	Africa	–	Vollesen, 1987#
G. capitis-viridis Mauer	Cape Verde Is.	B_3	Fryxell, 1979
G. triphyllum †† (Harvey & Sonder) Hochreutiner	Africa	B_2	Fryxell, 1979
Subsection *Pseudopambak* Prokh.:			
G. stocksii Masters in J.D. Hooker	Somalia, Oman, Pakistan	E_1	Fryxell, 1979
**G. areysianum* Deflers	Yemen	E_3	Fryxell, 1979
G. somalense (Gürke) J.B. Hutchinson	Africa	E_2	Fryxell, 1979
**G. bricchettii* (Ulbr.) Voll.	Somalia	–	Vollesen, 1987
**G. benadirense* Mattei	Africa	–	Vollesen, 1987
**G. incanum* (Schwartz) Hillcoat	Yemen	E_4	Fryxell, 1979
**G. sp.* A, sensu Voll.	Somalia	–	Vollesen, 1987#
Subsection *Longiloba* Fryx.:			
G. longicalyx J.B. Hutch. & Lee	Africa	F_1	Fryxell, 1979
Section *Serrata* Fryx.:			
**G. trifurcatum Voll.*	Somalia	–	Vollesen, 1987

Subgenus *Karpas* Rafinesque:

G. mustelinum Miers ex Watt	Brazil	AD_4	Fryxell, 1979
G. tomentosum Nuttall ex Seemann	Hawaiian Islands	AD_3	Fryxell, 1979
G. barbadense L.	--¶	AD_2	Fryxell, 1988
G. darwinii Watt	Galapagos Islands	–	Fryxell, 1979
G. hirsutum L.	--¶	AD_1	Fryxell, 1988
G. lanceolatum Tod.	Mexico	–	Fryxell, 1988

* Not available for distribution from the National Collection of *Gossypium* Germplasm.

† Binomial change from previously accepted designation, now proposed by Fryxell (pers. commun.).

‡ Any species that lacks published cytogenetic genomic analysis is left blank.

§ Although the G genomic designation was published, this designation is not now accepted by many cotton researchers.

¶ The cultivated cottons cannot be fully characterized for their original distribution. *G. arboreum* does not have any certainly known wild populations, and the remaining wild populations of *G. herbaceum* in Africa have been reclassified at a subspecies rank. The two tetraploids, *G. barbadense* and *G. hirsutum,* are found in wild stands throughout tropical regions of the western hemisphere, but are considered to have originated in South America and Mexico, respectively.

These species were listed by Vollesen (1987). Fryxell (pers. commun.) views *G. anomalum* subsp. *senarense* as warranting elevation to specific rank, and there is no formal designation yet of *Gossypium sp.* A., pending further study.

†† The placement of *G. triphyllum* is tentative in Subsection *Anomala* and has been included in Section *Hibiscoidea* (See: Fryxell, 1979).

Meredith (1991) mentions several later introductions which demonstrate that multiple, primitive strains contributed to early American cultivars. Meredith also documents that virtually all U.S. breeding programs rely on a foundation of plant introductions which included introgression from at least one *Gossypium* species other than *G. hirsutum.* The most well-known case, besides the obvious ubiquitous contribution of *G. barbadense* alleles, involves introgression from *G. arboreum* and *G. thurberi* with the development of a triple hybrid, *arboreum-thuberi-hirsutum,* by Beasley (1940). Meredith (1991) states that the triple hybrid germplasm became an important component of many modern cultivars. As plant breeding and genetics developed, the enhancement of cotton germplasm with introgression from wild species and races continued to be important. Several examples of successful introgression are given by Fryxell (1976) and Meredith (1991) that primarily involve disease resistance.

One of the most important problems for current cotton production is susceptibility to insects. Recent estimates of lint yield losses from insects nationwide in 1990 totaled $273 million, even though the average insecticide cost per acre is over $30 (Anonymous, 1991). Many germplasm enhancement programs, including the laboratory in College Station, TX, have utilized traditional methods of gene introgression from the germplasm resources to improve this situation. However, this methodology requires time to develop adapted lines that cotton breeders will accept in their hybridization programs. Meredith (1980) pointed out that once a breeder decides to use an insect resistant trait, from 8 to 12 years are required for the cotton breeder to release an improved variety after making the decision to cross with a germplasm line. Besides the time element to develop an initial introgressed breeding line, breeders will usually require a data collection period or impose delays due to their priorities before commencing the varietal improvement phase of development.

One example of successful introgression into Upland cotton is the absence of leaf and extrafloral nectaries (nectariless). A systematic program of interspecific cotton breeding began in Mississippi in 1952 (Meyer, 1974). Meyer and Meyer (1961) published the identification and transfer from *G. tomentosum* of the two recessive genes, ne_1 and ne_2, that condition the nectariless trait. Lukefahr and Rhyne (1960) and Lukefahr et al. (1965) demonstrated the resistance of nectariless cotton to several lepidopteran insects. A decade later, cotton geneticists were reporting the agronomic benefits of nectariless cottons (Meredith et al., 1973). The first important nectariless variety, Stoneville 825, was planted on about 10% of the U.S. cotton acreage from 1984 through 1987 (Meredith, 1991). However, despite continued reports of the usefulness of these genes (Wilson, 1989), commercial interest has waned for including this beneficial introgressed trait in routine cultivar development.

A leaf variant (okra-leaf), conditioned by the incompletely dominant allele $L_2{}^o$, also has been successfully utilized for insect resistance, although not in the U.S.A. on a large commercial scale. The okra-leaf variant was reported by Shoemaker (1908), and the allele was identified as derived from wild tetraploids (Stephens, 1945) later treated as *G. lanceolatum* by Fryxell (1979). The usefulness of okra-leaf for insect resistance has been documented with numerous reports (Jones, 1982). The Australian cotton breeding program at Narrabri, New South Wales incorporated this trait and three other characteristics into a Deltapine 61 background (Thomson et al., 1987). The cultivar Siokra was developed from this material which showed good resistance to *Heliothis* and, along with related releases, has constituted over

50% of total Australian acreage during the late 1980's (Thomson, pers. commun.).

Other traits from gene introgression programs have proven to confer moderate resistance to insects and could be cited as "successful" germplasm enhancement. In general, however, all of these materials have had limited utilization by commercial breeding companies. Meredith (1991) stated his frustration with the unrecognized value of moderately resistant cultivars and speculated that since no resistant trait from traditional gene introgression gave complete control without some insecticide use, the cotton community and the general public weren't motivated enough to accept these genetic lines and a reduced dependency on insecticides. This explanation could be correct when the enthusiastic reaction to transgenic plants containing the *Bt* gene (from the bacterium *Bacillus thuringiensis),* which almost completely controls a number of lepidopteran pests, is compared to the lukewarm response for many traditional gene introgression efforts.

GLANDING MODIFICATION: AN EXAMPLE OF PROBLEMATIC INTROGRESSION WITH TRADITIONAL BREEDING

Before examining the introgression results from biotechnology, I'll offer one additional example that can illustrate the complexity of gene exploitation for crop species such as cotton, which have numerous quality requirements for commercial cultivars. One of the unique botanical features of *Gossypium* and the other genera in the tribe Gossypieae is the presence of lysigenous glands throughout all above-ground tissues. Fryxell (1979) noted that these glands were one of two traits first used by Alefeld in 1861 to delineate the tribe. Later studies determined that these glands contained unique chemicals such as gossypol (Stanford and Viehoever, 1918). In the early 1960's, the chemicals in the glands were discovered to have important insecticidal properties (Lukefahr and Martin, 1966). A source of high frequency of glands was identified from a race accession of *G. hirsutum* from Socorro Island, Mexico, and several intensive breeding efforts were started to develop "high gossypol" germplasm.

Cotton germplasm with moderate levels of insect resistance has been released from several introgression programs. Lines incorporating the Socorro Island trait have been released from our project and breeding projects in Missouri, Louisiana, and Mississippi. The College Station USDA-ARS lines were the only materials to include additional high glanding sources from the *G. hirsutum* race collection (Altman et al., 1986). Most of this germplasm is not in a highly adapted genetic background, and only the Louisiana lines developed by Jack Jones compare favorably in yield and fiber properties to commercial check cultivars. However, once again a promising trait has not been accepted for development by commercial breeding companies.

In this instance, the reason for lack of acceptance of the introgressed germplasm is rather complicated. After gossypol was identified as an insecticidal compound, some researchers felt that gossypol was improperly identified as "the sole protector-chemical" (Chan et al:, 1988). Gossypol was assumed to be the major compound in all tissues, and methods for measuring seed gossypol were applied to various tissues subject to insect attack. We now know that there are five other major terpenoids in Upland cotton, that gossypol constitutes less than 10% of total terpenoids in important parts such as leaves, and that the standard seed analysis method cannot accurately quantitate terpenoids in other tissues (Stipanovic et al., 1988; Altman et al., 1989). Specifically targeted introgression was not considered in the original germplasm programs, and the "high gossypol" lines have higher terpenoid content in all tissues. The genetics of this trait is unknown, but we have measured high broad-sense heritabilities for all six Upland terpenoids (Altman et al., 1989). The high terpenoid characteristic from wild race stocks can be scored visually by measuring gland frequency in the calyx-crown area of flower buds. Although the gene(s) are unidentified that control this trait from exotic cottons, full expression can be obtained in advanced backcross segregants (Table 2).

Unfortunately, the general elevation of cotton terpenoids leads to some significant negative effects. Higher terpenoid seed is not desirable unless complete insect control is achieved. Cottonseed is the second largest oilseed crop worldwide (Cherry and Leffler, 1984) and worth $720 million annually in the U.S.A. (Anonymous, 1989). Gossypol is the major contaminant in cottonseed, and the cottonseed industry is reluctant to process seed with-higher gossypol levels.

In fact, the development of glandless cotton constitutes another gene introgression program with an emphasis on improving cottonseed quality. In first considering the concept of gland removal, this idea seems impossible since ubiquitous gland occurrence was considered as characteristic of not only the genus *Gossypium,* but the entire Gossypieae. However, McMichael (1960) discovered that complete glandlessness could be attained when two recessive alleles, gl_2 and gl_3, introgressed from the Hopi Indian cotton, were present. Kohel and Lee (1984) later determined that a single partially dominant allele, $Gl_2{}^e$, discovered by Egyptian researchers in *G. barbadense,* gives the same phenotype, when homozygous and introgressed into *G. hirsutum,* as does homozygous gl_2gl_3.

Before concluding that this parallel development would lead to another gene introgression success story, the insecticidal properties of pigment chemicals should not be forgotten. Glandless cotton has been reported to have greatly increased susceptibility to certain insects (Jenkins et al., 1986), and many cotton breeders have predicted very low production of glandless cotton (Niles and Feaster, 1984). The best of both worlds would be gossypol elimination in the seed, but an increase of glanding in other above-ground parts. This strategy of combining glandless seed with a glanded plant has been proposed (Marshall and Broué, 1981) because of the discovery of this characteristic in several wild species from Australia (Fryxell, 1965). Hexaploid hybrids between tetraploid *G. hirsutum* and the diploid Australian species *G. sturtianum* were derived as well as sterile pentaploids (Brown and Menzel, 1952; Muramoto, 1969). However, further gene introgression did not seem possible with traditional methodology because of the apparent remote relationship between these members of the genus.

One other strategy is being pursued in the glanding introgression programs. Following quantitation advances for the

Table 2. Backcross F_2 and parent calyx-crown gland counts† for four cycles of selection with the high calyx-glanding trait (BC_2 - BC_5).

Genotype	1988	1989	1990	1991
Tamcot CAMD-BC_xF_2				
highest segregant	153	136	132	168
average of visual selections	105	102	100	122
Stoneville 213 BC_xF_2				
highest segregant	126	116	110	135
average of visual selections	85	93	68	88
Parents				
ARSTX-HIGOS 3 (donor)	--	124	--	--
Tamcot CAMD-E (recurrent)	17	26	37	29
Stoneville 213 (recurrent)	8	10	13	11

† Counts were made on five calyx crown samples/individual plant, with all glands in the region included. For 1988-1990, total F_2 population sizes were about 400, and the top 20% were selected visually. The select 20% or approximately 60 plants were examined with a dissecting microscope. In 1991, F_2 populations were about 750, and the top 10% were selected visually. Ten individual plants were randomly sampled (five calyx-crown samples/plant) for parent lines.

six Upland cotton terpenoids, a thorough characterization of other races, species, and genera became possible. Terpenoids have been quantitated for A, D, and AD genome cottons (Altman et al., 1990b; Altman et al., 1991b). One of the D species, *G. raimondii*, produces a large amount of a unique compound, raimondal. Raimondal was discovered to be several-fold more toxic to tobacco budworm than any previously tested sesquiterpenoid (Stipanovic et al., 1990). The triploid hybrids and hexaploids produce all of the normal Upland terpenoids as well as substantial amounts of raimondal. Raimondal is not produced in seed of the parent diploid or of any hybrids and hexaploids, so targeted introgression for chemical profile changes in specific tissues seems to be a potential mechanism to increase insect resistance without reducing seed quality.

CELL CULTURE-ASSISTED INTROGRESSION

Cotton has not been on the forefront of developments in biotechnology until recently. In vitro culture of cotton ovules began in the early 1960's (Mauney, 1961). The first callus cultures were obtained from cultured ovules (Beasley, 1971), but mesocotyl (Schenk and Hildebrandt, 1972) and leaf disk (Davis et al., 1974) explants later were used successfully to culture callus. Although development of useful ovule culture technology and of totipotent callus systems was not rapid, both of these types of cell culture have made contributions to gene introgression.

Ovule culture, coupled with embryo culture, has been an effective method to obtain interspecific hybrids. There are numerous reports of interspecific crosses from different laboratories (Stewart and Hsu, 1978; Refaat et al., 1984; Gill and Bajaj, 1984; Umbeck and Stewart, 1985; Altman, 1988). Many of these crosses could have been obtained by standard crossing procedures, by standard procedures combined with exogenous hormone treatments, or by less rigorous culture methods relying on embryo rescue techniques. Yet, several hybrids have been produced with these methods which were previously considered to be unobtainable. Therefore, ovule and/or embryo culture allows a reduction of the barrier to gene transfer within the genus.

One example for gene introgression using ovule culture is the attempt to transfer the glandless seed and glanded plant trait from the wild Australian species, *G. sturtianum*. Our laboratory had received hexaploid (*hirsutum-sturtianum*) seed from H. Muramoto, but was unable to obtain backcross plants beyond pentaploids with this material. Using modified procedures from J.M. Stewart's studies (Stewart and Hsu, 1978; Umbeck and Stewart, 1985), we recovered backcross plants from pentaploids (Altman et al., 1987). One of these plants was missing glands on the seedling stem (hypocotyl equivalent tissue) and cotyledons, but developed normal glanding from the first true-leaf through maturity.

This plant became the parent in a long-term backcrossing program. We have definitively identified four monosomic addition lines (Rooney et al., 1991) and isolated a probable fifth (Altman, unpublished data). One of these lines, carrying the C_1-D chromosome, is a fully glanded plant which came from a glandless seed. Transmission data on the four confirmed monosomic additions have been reported (Rooney and Stelly, 1991). At present, besides the 53-chromosome (2n + 1) plants, we have approximately 500 normal germplasm lines that are being evaluated. Several of these lines combine greatly reduced seed glanding with normal glanding on other plant parts (Altman, unpublished data). However, complete expression of the desired phenotype has not yet been obtained in a 52-chromosome (2n) line.

Contributions to gene introgression from callus culture have received more attention. Cotton was categorized as a recalcitrant plant for achieving embryogenesis, and the first paper on regeneration from *G. hirsutum* callus was not published until 1983 (Davidonis and Hamilton, 1983). Several other groups followed with reports on embryogenic systems (Rangan et al., 1984; Mitten, 1985; Altman et al., 1985; Shoemaker et al., 1986; Trolinder and Goodin, 1987). Also of note, highly embryogenic cotton suspension cultures became feasible as well (Finer, 1988). Much information was not published because the bulk of the regeneration research was occurring in industrial laboratories that were restricted to various degrees from complete disclosures. We received for analysis the first regenerated cotton plants from Gayle Davidonis, and subsequently from most cotton tissue culture research programs. A high degree of cytogenetic aberrations was discovered (Stelly et al., 1989), as well as subtle quantitative genetic changes (Altman et al., 1991a).

Following these breakthroughs for totipotent cotton cell cultures, several groups reported successful transformation using both the *Agrobacterium* system (Umbeck et al., 1987) and the particle gun (Finer and McMullen, 1990). The particulars of transformation methodologies have been reviewed previously (Gasser and Fraley, 1989), but generally the literature contains little information about gene action following transformation, except for a few reports with marker genes. Initial data on cotton foreign gene inheritance have examined genes for kanamycin and chloramphenicol resistance (Umbeck et al., 1989). Paul Umbeck's group concluded that single gene insertions were inherited and expressed in typical Mendelian fashion.

However, I would propose an alternative hypothesis that foreign genes will not behave always as Mendelian traits. Because preferential chromatin transmission is found with other gene introgression from exotic germplasm (Rooney and Stelly, 1991) and several characterized qualitative traits do not exhibit absolute Mendelian inheritance in cotton, foreign genes could demonstrate the same phenomena. We have transferred two foreign gene constructions into 28 different populations in three isoline groups (Sachs et al., unpublished data). F_2 data from eight of the 56 populations are typical of the trends (Table 3). For certain population-foreign gene combinations, such as the two Tamcot CAMD-E populations with HD73, a normal 3:1 segregation was observed. Yet, the HDl gene in the same genetic backgrounds did not fit a 3:1 pattern. I would suggest that whatever inheritance and expression model proves to be accurate, biotechnologists should not assume that foreign gene genetics will be straightforward.

Cotton should provide many opportunities to experiment with transgenic plants Due to the economic incentive for developing insect-resistant cotton, the availability of characterized genes with a good potential for effective and possibly complete control of the major pests, and the perception that cotton is primarily a fiber crop and not a food commodity, transgenic cottons were produced with gene constructions coding for a modified δ-endotoxin from *Bacillus thuringiensis* var. *kurstaki (Bt* gene). Paul Umbeck's group from Agracetus conducted the first field trials in cooperation with USDA-ARS at Mississippi State, MS (Umbeck et al., 1990). We also examined insect behavior and plant injury with these cottons in greenhouse experiments (Benedict et al., 1991; Benedict et al., in press). Results from these experiments suggested some effects of the foreign gene, but not enough to be of economic importance.

Monsanto was ready in 1990 for field-testing other transgenic cottons with the *Bt* gene developed by Randy Deaton's group. Researchers at Monsanto created constructions that exhibited up to 500-fold increase in expression of the *Bt* gene by further modifying the gene to more closely resemble a functional plant coding sequence (Perlak et al., 1990). Monsanto in cooperation with our laboratory and other USDA-ARS and state experiment stations used seven locations to test transgenic cottons across the U.S.A., and preliminary results have been published recently (Deaton, 1991; and six following abstracts in the same volume). Results showed that these cottons confer a high degree of resistance to lepidopteran insects.

Other kinds of transgenic cottons also are being developed. Calgene has conducted field tests of transgenic cottons with bromoxynil (Buctril®) herbicide resistance (Kiser and Mitchell, 1991). Monsanto plans to field-test another transgenic cotton with glyphosate (Round-up®) resistance. A preliminary report indicated development of 2,4-D resistant transgenic cotton

Table 3. Transmission of foreign genes† in progeny from crosses with transgenic cotton.

Genetic Background‡	Foreign Gene	Number of Plants§ (+)	(-)	Total	c^2 ¶	P¶
Tamcot CAMD-E						
glandless	pMON5377	53	4	57		
normal	pMON5377	19	4	23		
pooled HDl	pMON5377	72	8	80	8.82	0.005-0.001
glandless	pMON5383	21	9	30		
normal	pMON5383	48	18	66		
pooled HD73	pMON5383	69	27	96	0.35	0.75-0.50
Stoneville 213						
glandless	pMON5377	24	4	28		
normal	pMON5377	55	14	69		
pooled HDl	pMON5377	79	18	97	2.07	0.25-0.10
glandless	pMON5383	80	23	103		
normal	pMON5383	91	20	111		
pooled HD73	pMON5383	171	43	214	2.49	0.25-0.10

† The two foreign gene constructions (developed by Monsanto Company, St. Louis, MO) code for the modified and truncated d-endotoxin proteins from *Bacillus thuringiensis* var. *kurstaki,* races HDl and HD73, and are designated as pMON5377 and pMON5383, respectively, in the table.

‡ These F_2 populations are from crosses between BC_4 isolines of highly inbred (15+ generations of selfing) cultivar selections, Tamcot CAMD-E and Stoneville 213, and an R_3 (explant source cultivar: Coker 312) transgenic plant expressing the foreign gene. The glandless characteristic is controlled by the Gl_2^e allele.

§ Plants were rated by an ELISA assay (developed by Monsanto Company, St. Louis, MO) for expression (+) or non-expression (-) of the foreign gene.

¶ Chi-square values (c^2) and corresponding probability levels (P) are for testing a proposed ratio of 3:1 for all populations.

(Trolinder et al., 1991). An important consideration is that all of the advanced transformed cottons have come from explants from the Coker germplasm, leading to concerns about the adaptability of these lines, the narrow gene base, and the genotypic specificity of any somaclonal variation.

Problems, or opportunities in certain situations, arising from somaclonal variation must be carefully considered, given the rapid development of transgenic cottons. Although cotton is no longer viewed as too recalcitrant to permit practical gene introgression via biotechnology, the current culture systems have several limitations. Norma Trolinder's group has published some of the most extensive genotype screening for embryogenic response and has found that efficient regeneration is limited to the Coker germplasm (Trolinder and Xhixian, 1989). The time frame is considerably longer from explant to fruiting regenerant than with most other crop culture systems. As indicated above, the degree of somaclonal variation, within the constraints of our current cell culture technology, is relatively high. Gould et al. (1991) recently reported that adaptations of the previous shoot tip culture system could provide an alternative without the intrinsic genotypic limitations of present callus regeneration technology.

Given the extensive use of cotton transformation with somatic embryos, scientists will need to conduct experiments to separate somaclonal variation effects from potential insertion positional effects. Our agronomic data from some of the first field trials of transgenic cottons show that somaclones carrying the same gene insertion differ for most trait measurements (Table 4). Future experimentation might clarify the mechanism controlling these responses, but from our initial studies (Altman et al., 1991a), we would hypothesize that quantitative differences result from inherent somaclonal variation due to culture systems, genotype specificity, a genotypic bottleneck from the regenerative cells per se, and/or other unknown factors.

If principles from traditional plant breeding are applied to transgenic plant development using the available data, more efficient cotton improvement strategies could be formulated. Based on our experience and data, we have proposed a breeding strategy for gene introgression using regenerated cotton plants (Figure 1). We regard this proposal as a subject for discussion and not a conclusive methodology. However, the important factors that any cotton improvement programmust address are: (1) how to avoid cryptic variation from cytogenetic or genetic origins, (2) how to maximize genetic gain given somaclonal effects on genetic variation, (3) how to utilize efficient early generation selection to improve the likelihood of obtaining acceptable levels of the numerous quality and yield parameters for a commercial cultivar, and (4) how to balance rigorous science with economic pressures to release products for commercialization.

Enthusiastic proponents of biotechnology should recognize that the long varietal development time (8-12 years by Meredith's 1980 estimate) will still be required with gene introgression germplasm from new technologies. Given the rapid results from cotton transformation already, the time element for production of introgressed germplasm probably can be shortened. Yet, the

Table 4. Yield trial means of agronomic traits[†] for transgenic lines[‡] and check cultivars at College Station, TX in 1990.

Genotype	Lint Yield[§] (#/A)	Lint %	Earliness (%)	Boll Weight (gm)	Micronaire	Staple Length (inches)	Uniformity (%)	Fiber Strength (cN/tex)	E1%
Coker 312	817	37.0	50	5.20	3.9	1.18	44.6	19.0	7.2
MON 81	1,076	37.1	46	4.07	4.4	1.12	47.6	19.6	6.8
MON 62	1,039	36.8	29	4.37	4.3	1.15	48.5	19.5	7.3
MON 65	1,199	35.9	54	4.37	3.8	1.14	47.8	19.5	7.4
MON 82	1,147	38.1	57	4.30	3.4	1.15	47.6	20.8	7.2
MON 84	1,160	36.5	48	4.41	4.0	1.17	46.7	18.9	7.5
Stoneville 453	990	39.0	55	5.32	4.2	1.16	47.7	18.5	8.0
$LSD_{(.05)}$	161	1.3	14	0.27	0.3	0.02	2.0	0.7	0.4

† Means for lint yield, lint percentage, earliness, and boll weight were from subplots grown without lepidopteran control; fiber measurements were from subplots receiving standard insecticide applications.

‡ Transgenic lines were R_3 seed (explant source cultivar: Coker 312) expressing a modified and truncated δ-endotoxin protein from *Bacillus thuringiensis* var. *kurstaki* race HD1 and developed by Monsanto Company, St. Louis, MO. All transgenic lines have the designation MON.

§ Stoneville 453 yielded significantly more lint ($P < 0.05$) than the transgenic lines and Coker 312 with standard insecticide protection. Yield reductions in comparing these lint yields with corresponding subplots protected against lepidopteran damage were 37, 42, and 18% for Coker 312, Stoneville 453, and the average of the MON lines, respectively.

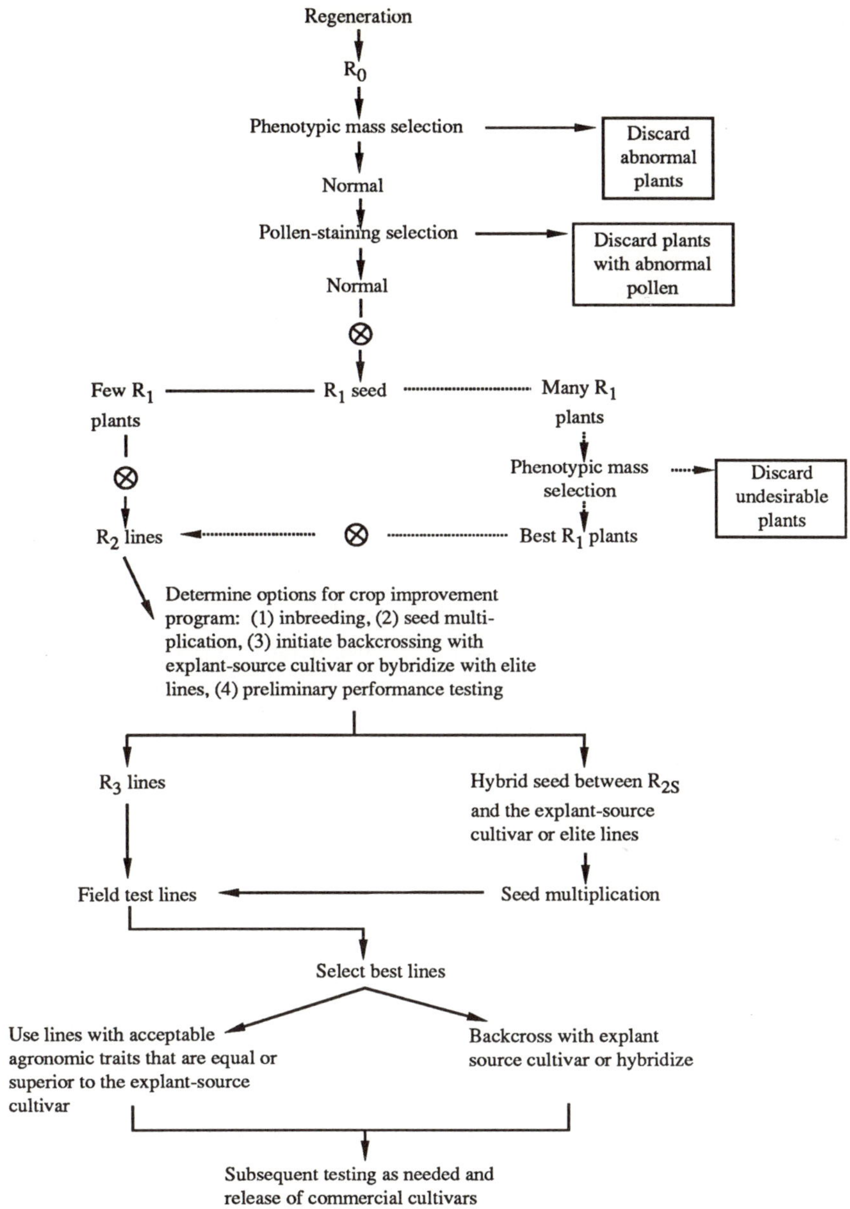

Fig. 1. Proposed cotton regeneration and somaclonal development plan (Altman et al., 1991a). Dashed lines indicate an option as mass extraction of normal cotton regenerants becomes available.

cotton breeding step will remain the same or could be longer, since somaclonal effects dictate modified procedures, unfamiliar gene measurement techniques might be necessary, and regulatory guidelines could force delays.

CONCLUSIONS

Gene introgression in cotton has a long history. The genetic resources within the genus *Gossypium* have been the foundation for useful introgression. The germplasm available in nature is represented incompletely in collections. A major commitment for genetic conservation, coupled with continued taxonomic research, is essential to insure genetic enhancement. The value of gene introgression should not be underestimated. However, successful utilization of this process for improving cotton production, unfortunately, is not obtainable in as short of a time frame as we might wish.Biotechnologies have broadened the possibilities of gene transfer for cotton. Specific applications, such as attaining a high level of insect resistance with the *Bt* gene insertions, seem to be commercially possible within the next few years, from the scientific point of view although not necessarily from a regulatory perspective. I believe that many of the gene introgression programs using traditional methodology will have as much to offer as the new biotechnologies. Biotechnological gene introgression is just another tool for practical cotton improvement, albeit a tool that could eliminate many biological barriers to gene introgression. But, rigorous investigations of scientific potential and careful development of transgenic cottons will be the most cost-efficient in the long term. A hasty deployment or a tendency to oversell the technology will only set back actual utilization and acceptance by the farming community. We should not draw too many conclusions without having adequate data in hand.

ACKNOWLEDGMENTS

I want to thank Drs. J.H. Benedict, W.R. Deaton, P.A. Fryxell, R.J. Kohel, W.R. Meredith Jr., A.E. Percival, and F.D. Wilson for critically reviewing the manuscript. Mention of a trademark or proprietary product does not constitute a guarantee or warranty of the product by the USDA and does not imply its approval to the exclusion of others that may also be suitable.

REFERENCES

Altman, D.W., 1988, Exogenous hormone applications at pollination for in vitro and in vivo production of cotton interspecific hybrids, *Plant Cell Reports* 7:257-261.

Altman, D.W., Fryxell, P.A., Koch, S.D., and Howell, C.R., 1990a, *Gossypium* germplasm conservation augmented by tissue culture techniques for field collecting, *Econ. Bot.* 44:106-113.

Altman, D.W., Stelly, D.M., and Kohel, R.J., 1987, Introgression of the glanded-plant and glandless-seed trait from *Gossypium sturtianum* Willis into cultivated Upland cotton using ovule culture, *Crop Sci.* 27:880-884.

Altman, D.W., Stelly, D.M., and Mitten, D.M., 1991a, Quantitative trait variation in phenotypically normal regenerants of cotton, *In Vitro Cell. Devel. Biol.* 27P: 132-138.

Altman, D.W., Stipanovic, R.D., and Bell, A.A. 1990b, Terpenoids in foliar pigment glands of A, D, and AD genome cottons: introgression potential for pest resistance, *J. Hered.* 81:447-454.

Altman, D.W., Stipanovic, R.D., and Bell, A.A., 1991b, Terpenoids of Asiatic and Western Hemisphere diploid and tetraploid cottons, *in* "Proc. Beltwide Cotton Conf., vol. 1," D.J. Herber, ed., San Antonio, TX. 8-12 Jan. 1991. National Cotton Council of America, Memphis, TN, p. 534-537.

Altman, D.W., Stipanovic, R.D., and Benedict, J.H., 1986, *Annual Plant Resistance Insects Newsl.* 12:32-33.

Altman, D.W., Stipanovic, R.D., and Benedict, J.H., 1989, Terpenoid aldehydes in Upland cottons. II. Genotype-environment interactions, *Crop Sci.* 29:1451-1456.

Altman, D.W., Stipanovic, R.D., Mitten, D.M., and Heinstein, P., 1985, Interaction of cotton tissue culture cells and *Verticillium dahliae, In Vitro Cell. Devel. Biol.* 21:659-664.

Anonymous, 1989, Agricultural Statistics 1989, U.S. Government Printing Office, Washington, D.C., 547 p.

Anonymous, 1991, *Cotton Management Guide, Southwestern Edition* 8:1.

Barrow, J.R., 1986, The conditions required to isolate and maintain viable cotton *(Gossypium hirsutum* L.) microspores, *Plant Cell Reports* 5:405-408.

Beasley, C.A., 1971, In vitro culture of fertilized cotton ovules, *BioScience* 21:906-907.

Beasley, J.O., 1940, The production of polyploids in *Gossypium, J. Hered.* 31:39-48.

Benedict, J.H., Altman, D.W., Umbeck, P.F., and Ring, D.R., 1991, Patterns of behavior, growth, survival and plant injury for *Heliothis virescens* (F.) (Lepidoptera: Noctuidae) on transgenic *Bt* cottons, Texas Agric. Exp. Stn. Progress Rep. #4814. 12 p.

Benedict, J.H., Altman, D.W., Umbeck, P.F., and Ring, D.R., 1991, Patterns of behavior, growth, survival and plant injury for *Heliothis virescens* (F.) (Lepidoptera: Noctuidae) on transgenic *Bt* cottons, *J. Econ. Entomol.* (In press)

Brown, M.S., and Menzel, M.Y., 1952, Polygenomic hybrids in *Gossypium.* I. Cytology of hexaploids, pentaploids and hexaploid combinations, *Genetics* 37:242-263.

Chan, B.G., Mahoney, N., and Wilson, F.D., 1988, A holistic approach to study HPR in cotton, *in* "Proc. Beltwide Cotton Production Res. Conf.", J.M. Brown, ed., New Orleans, LA. 3-8 Jan. 1988. National Cotton Council of America, Memphis, TN, p. 100-105.

Cherry, J.P., and Leffler, H.R., 1984, Seed, *in* "Cotton," R.J. Kohel and C.F. Lewis, eds., American Society of Agronomy, Madison, WI, p. 511-569.

Davidonis, G.H., and Hamilton, R.H., 1983, Plant regeneration from callus tissue of *Gossypium hirsutum* L., *Plant Sci. Letters* 32:89-93.

Davis, D.G., Dusbabek, K.E., and Hoerauf, R.A., 1974, In vitro culture of callus tissues and cell suspensions from okra *(Hibiscus esculentus* L.) and cotton *(Gossypium hirsutum* L.), *In Vitro* 9: 395-398.

Deaton, W.R., 1991, Field performance of cotton genetically modified to express insecticidal protein from *Bacillus thuringiensis* I. Introduction, *in* "Proc. Beltwide Cotton Conf., vol. 1," D.J. Herber, ed., San Antonio, TX. 8-12 Jan. 1991. National Cotton Council of America, Memphis, TN., p. 576.

Finer, J.J., 1988, Plant regeneration from somatic embryogenic suspension cultures of cotton *(Gossypium hirsutum* L.), *Plant Cell Reports* 7:399-402.

Finer, J.J., and McMullen, M.D., 1990, Transformation of cotton *(Gossypium hirsutum* L.) via particle bombardment, *Plant Cell Reports* 8:586-589.

Fryxell, P.A., 1965, A revision of the Australian species of *Gossypium* with observations on the occurrence of *Thespesia* in Australia (Malvaceae), *Austral. J. Bot.* 13:71-102.

Fryxell, P.A., 1968, The typification and application of the Linnaean binomials in *Gossypium, Brittonia* 20:378-386.

Fryxell, P.A., 1976, Germpool utilization: *Gossypium,* a case history, USDA Rep. ARS-S-137. 7 p.

Fryxell, P.A., 1979, The natural history of the cotton tribe, Texas A&M University Press, College Station, TX. 245 p.

Fryxell, P.A., 1988, Malvaceae of Mexico. *Systematic Bot. Monographs* 25:1-522.

Fryxell, P.A., Craven, L.A., and Stewart, J.M., 1992, A revision of *Gossypium* Section *Grandicalyx* (Malvaceae), including the description of six new species, *Systematic Bot.* 17. (In press)

Gasser, C.S., and Fraley, R.T., 1989, Genetically engineered plants for crop improvement, *Science* 244: 1293-1299.

Gill, M.S., and Bajaj, Y.P.S., 1984, Interspecific hybridization in the genus *Gossypium* through embryo culture, *Euphytica* 33:305-311.

Gould, J., Banister, S., Hasegawa, O., Fahima, M., and Smith, R.H., 1991, Regeneration of *Gossypium hirsutum* and *G. barbadense* from shoot apex tissues for transformation, *Plant Cell Reports* 10: 12-16.

Hutchinson, J.(B)., 1959, The application of genetics to cotton improvement, Cambridge University Press, London. 87 p.

Jenkins, J.N., Parrott, W.L., McCarty, J.C., and Dearing, L., 1986, Performance of cottons when infested with tobacco budworm, *Crop Sci.* 26:93-95.

Jones, J.E., 1982, The present state of the art and science of cotton breeding for leaf-morphological types, *in* "Proc. Beltwide Cotton Production Res. Conf.", J.M. Brown, ed., Las Vegas, NV. 37 Jan. 1982. The National Cotton Council of America, Memphis, TN., p. 93-99.

Jones, V.H., 1936, A summary of data on aboriginal cotton of the Southwest, *in* "Symposium on Prehistoric Agriculture," Univ. New Mexico Bull. #296, Anthropol. Ser., vol. 1, no. 5. Las Cruses, NM., p. 51-64.

Kiser, J., and Mitchell, J.F., 1991, Bromoxynil safened cotton: trialing history, yield and quality, *in* "Proc. Beltwide Cotton Conf., vol. 1," D.J. Herber, ed., San Antonio, TX. 8-12 Jan. 1991. National Cotton Council of America, Memphis, TN., p. 566-568.

Kohel, R.J., and Lee, J.A., 1984, Genetic analysis of Egyptian glandless cotton, *Crop Sci.* 24:1119-1121.

Lee, J.A., 1984, Cotton as a world crop, *in* "Cotton," R.J. Kohel and C.F. Lewis, eds., American Society of Agronomy, Madison, WI., p. 1-25.

Lewton, F.L., 1912, The cotton of the Hopi Indians; a new species of *Gossypium,* Smithsonian Misc. Coll. 60, No. 6, Pub. 2146. 10 p.

Lukefahr, M.J., and Martin, D.F., 1966, Cotton-plant pigments as a source of resistance to the bollworm and tobacco budworm, *J. Econ. Entomol.* 59:176-179.

Lukefahr, M.J., Martin, D.F., and Meyer, J.R., 1965, Plant resistance to five Lepidoptera attacking cotton, *J. Econ. Entomol.* 58:516-518.

Lukefahr, M.J., and Rhyne, C., 1960, Effects of nectariless cottons on populations of three lepidopterous insects, *J. Econ. Entomol.* 53:242-244.

Maréchal, R., 1983, Une collection d'hybrides interspécifiques du genre *Gossypium, Cot. Fib. Trop.* 38:240-246.

Marshall, D.R., and Broué, P., 1981, The wild relatives of crop plants indigenous to Australia and their use in plantbreeding,*J. Austral. Institute Agric. Sci.* 47:149-154.

Mauney, J.R., 1961, The culture in vitro of immature cotton embryos, *Bot. Gaz.* 122:205-209.

McCarty, J.C., Jenkins, J.N., Parrott, W.L., and Creech, R.G., 1979, The conversion of photoperiodic primitive race stocks of cotton to day-neutral stocks, Mississippi Agric. Forestry Exp. Stn. Res. Rep., vol. 4, no. 19. 4 p.

McMichael, S.C., 1960, Combined effects of glandless genes gl_2 and gl_3 on pigment glands in the cotton plant, *Agron. J.* 52:385-386.

Meredith, W.R., Jr., 1980, Use of insect resistant germplasm in reducing the cost of production in the 1980's, *in* "Proc. Beltwide Cotton Production Res. Conf.", J.M. Brown, ed., St. Louis, MO. 610 Jan. 1980. The National Cotton Council of America, Memphis, TN., p. 307-310.

Meredith, W.R., Jr., 1991, Contributions of introductions to cotton improvement, in "Use of plant introductions in cultivar development, Part 1," H.L. Shands and L.E. Wiesner, eds., Crop Science Society of America, Madison, WI., p. 127-146.

Meredith, W.R., Jr., Laster, M.L., Ranney, C.D., and Bridge, R.R., 1973, Agronomic potential of nectariless cotton *(Gossypium hirsutum* L.), *J. Environ. Qual.* 2:141-144.

Meyer, J.R., and Meyer, V.G., 1961, Origin and inheritance of nectariless cotton, *Crop Sci.* 1:167-169.

Meyer, V.G., 1974, Interspecific cotton breeding, *Econ. Bot.* 28:56-60.

Mitten, D.M., 1985, Somatic embryogenesis in *Gossypium hirsutum* L., *in* "Proc. Beltwide Cotton Production Res. Conf.", T.C. Nelson, ed., New Orleans, LA. 6-11 Jan. 1988. The National Cotton Council of America, Memphis, TN., p. 57-58.

Muramoto, H., 1969, Hexaploid cotton: some plant and fiber properties, *Crop Sci.* 26:27-29.

Narayanan, S.S., Singh, J., and Varma, P.K., 1984, Introgressive gene transfer in *Gossypium.* Goals, problems, strategies and achievements, *Cot. Fib. Trop.* 39:123-135.

Niles, G.A., and Feaster. C.V., 1984, Breeding, *in* "Cotton," R.J. Kohel and C.F. Lewis, eds., American Society of Agronomy, Madison, WI., p. 201-231.

Percival, A.E., 1987, The National Collection of *Gossypium* Germplasm, South. Coop. Ser. Bull. #321. 362 p.

Percival, A.E., and Kohel, R.J., 1990, Distribution, collection, and evaluation of *Gossypium, Adv. Agron.* 44:225-256.

Percy, R.G., and Turcotte, E.L., 1989, Bacterial blight resistance screen of *Gossypium barbadense* L. germplasm and the day-neutral conversion status of blight resistant accessions, *in* "Proc. Beltwide Cotton Production Res. Conf.", J.M. Brown, ed., Nashville, TN 2-7 Jan 1989. The National Cotton Council of America, Memphis, TN., p. 38-40.

Percy, R.G., and Wendel, J.F., 1990, Allozyme evidence for the origin and diversification of *Gossypium barbadense* L., *Theor. Appl. Genet.* 79:529-542.

Perlak, F.J., Deaton, R.W., Armstrong, T.A., Fuchs, R.L. Sims, S.R., Greenplate, J.T., and Fischhoff, D.A., 1990, Insect resistant cotton plants, *Bio/Technology* 8:939-943.

Rangan, T.S., Zavala, T., and Ip. A., 1984, Somatic embryogenesis in tissue cultures of *Gossypium hirsutum* L., *In Vitro* 20:256.

Refaat, M., Rossignol, L., and Demarly, Y., 1984, Interspecific hybrid *Gossypium hirsutum* L. x *G. barbadense* L. via in vitro fertilization and ovule culture, *Z. Pflanzenzuchtg.* 93:137-146.

Renfroe, M.H., and Smith, R.H., 1986, Cotton shoot tip culture, in "Proc. Beltwide Cotton Production Res. Conf.", T.C. Nelson, ed., Las Vegas, NV. 5-9 Jan 1986. National Cotton Council of America, Memphis, TN., p. 78-79.

Rooney, W.L., and Stelly, D.M., 1991, Preferential transmission and somatic elimination of a *Gossypium sturtianum* chromosome in *G. hirsutum, J. Hered.* 82:151-155.

Rooney, W.L., Stelly, D.M., and Altman, D.W., 1991, Identification of four *Gossypium sturtianum* monosomic alien addition derivatives from a backcrossing program with *G. hirsutum, Crop Sci.* 31:337-341.

Schenk, R.U., and Hildebrandt, A.C., 1972, Medium and techniques for induction and growth of monocotyledonous and dicotyledonous plant cell cultures, *Can. J. Bot.* 50:199-204.

Shoemaker, D.N., 1908, A study of leaf characters in cotton hybrids, *Rep. Am. Breed. Assoc.* 5:1161 19.

Shoemaker, R.C., Couche, L.J., and Galbraith, D.W., 1986, Characterization of somatic embryogenesis and plant regeneration in cotton (*Gossypium hirsutum* L.), *Plant Cell Reports* 3:178-181.

Smith, C.E., Jr., and Stephens, S.G., 1971, Critical identification of Mexican archaeological cotton remains, *Econ. Bot.* 25:160-168.

Stanford, E.E., and Viehoever, A., 1918, Chemistry and histology of the glands of the cotton plant, with notes on the occurrence of similar glands in related plants, *J. Agric. Res.* 13:419-435.

Stelly, D.M., Altman, D.W., Kohel, R.J., Rangan, T.S., and Commiskey E., 1989, Cytogenetic abnormalities of cotton somaclones from callus cultures, *Genome* 32:762-770.

Stephens, S.G., 1945, A genetic survey of leaf shape in New World cottons---a problem in critical identification of alleles, *J. Genet.* 46:313-330.

Stephens, S.G., and Moseley, M.E., 1974, Early domesticated cottons from archaeological sites in central coastal Peru, *Am. Antiquity* 39:109-122.

Stewart, J.M., and Hsu, C.L., 1978, Hybridization of diploid and tetraploid cottons through in-ovulo embryo culture, *J. Hered.* 69:404-408.

Stipanovic, R.D., Altman, D.W., Begin, D.L., Greenblatt, G.A., and Benedict, J.H., 1988, Terpenoid aldehydes in Upland cottons: analysis by aniline and HPLC methods, *J. Agric. Food Chem.* 36:509-5 15.

Stipanovic, R.D., Elissalde, M.H., Altman, D.W., and Norman, J.O., 1990, Cell culture bioassay to evaluate allelochemical toxicity to *Heliothis virescens* (Lepidoptera: Noctuidae), *J. Econ. Entomol.* 83:737-741.

Thomson, N.J., Reid, P.E., and Williams, E.R., 1987, Effects of the okra leaf, nectariless, frego bract and glabrous conditions on yield and quality of cotton lines, *Euphytica* 36:545-553.

Trolinder, N., Bailey, C., Quisenberry, J., and Ow, D., 1991, 2,4-D resistant transgenic cotton, *In Vitro Cell. Devel. Biol.* 27:95A.

Trolinder, N.L., and Goodin, J.R., 1987, Somatic embryogenesis and plant regeneration of cotton *(Gossypium hirsutum* L.), *Plant Cell Reports* 6:231-234.

Trolinder, N.L., and Xhixian, C., 1989, Genotype specificity of the somatic embryogenesis response in cotton, *Plant Cell Reports* 8:133-136.

Umbeck, P., Johnson, G., Barton, K., and Swain, W., 1987, Genetically transformed cotton *(Gossypium hirsutum* L.) plants, *Bio/Technology* 5:263-266.

Umbeck, P., Swain, W., and Yang, N.-S., 1989, Inheritance and expression of genes for kanamycin and chloramphenicol resistance in transgenic cotton plants, *Crop Sci.* 29:196-201.

Umbeck, P.F., Barton, K.A., Jenkins, J.N., and Parrott, W.L., 1990, Transgenic cotton plants: field test preparations and new developments in insecticidal proteins, *in* "Proc. Beltwide Cotton Production Res. Conf.", J.M. Brown, ed., Las Vegas, NV. 9-14 Jan. 1990. National Cotton Council of America, Memphis, TN., p. 635.

Umbeck, P.F., and Stewart, J.M., 1985, Substitution of cotton cytoplasms from wild diploid species for cotton germplasm improvement, *Crop Sci.* 25:1015-1019.

Vollesen, K., 1987, The native species of *Gossypium* (Malvaceae) in Africa, Arabia, and Pakistan, *Kew Bull.* 42:337-349.

Watt, G., 1907, The wild and cultivated cotton plants of the world, Longmans, Green, & Co., London. 406 p.

Wilson, F.D., 1989, Yield, earliness, and fiber properties of cotton carrying combined traits for pink bollworm resistance, *Crop Sci.* 29:7-12.

INDEX